United States Nuclear Regulatory Commission

Protecting People and the Environment

NUREG-2105, Vol. 4

I0493576

Environmental Impact Statement for the Combined License (COL) for Enrico Fermi Unit 3

Final Report

Appendices F to M

U.S. Nuclear Regulatory Commission
Office of New Reactors
Washington, DC 20555-0001

Regulatory Office
Permit Evaluation, Eastern Branch
U.S. Army Engineer District, Detroit
U.S. Army Corps of Engineers
Detroit, MI 48226

**US Army Corps
of Engineers**

AVAILABILITY OF REFERENCE MATERIALS
IN NRC PUBLICATIONS

United States Nuclear Regulatory Commission

Protecting People and the Environment

NUREG-2105, Vol. 4

Environmental Impact Statement for the Combined License (COL) for Enrico Fermi Unit 3

Final Report

Appendices F to M

Manuscript Completed: November 2012
Date Published: January 2013

U.S. Nuclear Regulatory Commission
Office of New Reactors
Washington, DC 20555-0001

Regulatory Office
Permit Evaluation, Eastern Branch
U.S. Army Engineer District, Detroit
U.S. Army Corps of Engineers
Detroit, MI 48226

**US Army Corps
of Engineers**

Final Environmental Impact Statement
for Combined License (COL) for Enrico Fermi Unit 3

Lead Agency:	U.S. Nuclear Regulatory Commission
Cooperating Agency:	Department of the Army U.S. Army Corps of Engineers, Detroit District
Contact:	Bruce Olson, Project Manager Environmental Projects Branch 2 Division of New Reactor Licensing Office of New Reactors U.S. Nuclear Regulatory Commission Washington, DC 20555-0001 phone: 301-415-3731 e-mail: Bruce.Olson@nrc.gov

Abstract:

This environmental impact statement (EIS) has been prepared in response to an application submitted to the U.S. Nuclear Regulatory Commission (NRC) by Detroit Edison for a construction permit and operating license (combined license or COL). The proposed actions related to the Detroit Edison application are (1) NRC issuance of a COL for a new power reactor unit at the Detroit Edison Enrico Fermi Atomic Power Plant (Fermi) site in Monroe County, Michigan; and (2) U.S. Army Corps of Engineers (USACE) permit action to perform certain regulated activities on the site. The USACE is participating with the NRC in preparing this EIS as a cooperating agency and participates collaboratively on the review team.

This EIS includes the NRC staff's analysis, which considers and weighs the environmental impacts of constructing and operating a new nuclear unit at the Fermi site and at alternative sites, and mitigation measures available for reducing or avoiding adverse impacts. Based on its analysis, the staff determined that there are no environmentally preferable or obviously superior sites.

The EIS includes the evaluation, in part, of the proposed action's impacts on the public interest, including impacts on waters of the United States pursuant to Section 404 of the Clean Water Act and Section 10 of the Rivers and Harbors Appropriations Act of 1899. The USACE will decide whether to issue a permit on the basis of the EIS evaluation of the probable impacts on the public interest, including cumulative impacts, of Detroit Edison's proposed activities that are within the USACE scope of analysis; USACE verification of compliance with the requirements of USACE regulations and the Clean Water Act Section 404(b)(1) Guidelines; and any supplemental information, evaluations, or verifications that may be outside the NRC's scope of analysis and not included in this EIS, but are required by the USACE to support its permit decision.

After considering the environmental aspects of the proposed action, the staff's recommendation to the Commission is that the COL be issued as proposed.[a] This recommendation is based on (1) the application, including the Environmental Report (ER) submitted by Detroit Edison; (2) consultation with Federal, State, Tribal, and local agencies; (3) the staff's independent review; (4) the staff's consideration of comments related to the environmental review that were received during the public scoping process

---

(a) As directed by the Commission in CLI-12-16, the NRC will not issue the COL prior to completion of the ongoing rulemaking to update the Waste Confidence Decision and Rule (see Section 6.1.6 of this EIS).

and on the draft EIS; and (5) the assessments summarized in this EIS, including the potential mitigation measures identified in the ER and this EIS. The USACE permit decision would be made following issuance of this final EIS and completion of its permit application review process and permit decision documentation.

Contents

Figures

Tables

Executive Summary

By letter dated September 18, 2008, the U.S. Nuclear Regulatory Commission (NRC or the Commission) received an application from Detroit Edison Company (Detroit Edison) for a combined license (COL) for a new power reactor unit, the Enrico Fermi Unit 3 (Fermi 3), at the Detroit Edison Enrico Fermi Atomic Power Plant (Fermi) site in Monroe County, Michigan.

The proposed actions related to the Fermi 3 application are (1) NRC issuance of COLs for construction and operation of a new nuclear unit at the Fermi site and (2) U.S. Army Corps of Engineers (USACE) permit action pursuant to Section 404 of the Federal Water Pollution Control Act, as amended (33 USC 1251, *et seq.*) (Clean Water Act), and Section 10 of the Rivers and Harbors Appropriation Act of 1899 (33 USC 403 *et seq.*) (Rivers and Harbors Act of 1899) to perform certain regulated activities associated with the Fermi 3 project, within the USACE jurisdiction and scope of analysis. The USACE is participating with the NRC in preparing this environmental impact statement (EIS) as a cooperating agency and participates collaboratively on the review team. The reactor specified in the application is an Economic Simplified Boiling Water Reactor (ESBWR) designed by GE-Hitachi Nuclear Energy Americas, LLC (GEH). The GEH design was approved by the NRC in March 2011. The final design approval was published in the *Federal Register* on March 16, 2011 (76 FR 14437).

The NRC staff completed its safety review of the ESBWR design on March 9, 2011 and issued a final safety evaluation report (FSER, Agencywide Documents Access and Management System [ADAMS] accession number ML103470210). The NRC staff also issued a standard design approval (SDA) via letter to GE Hitachi Nuclear Energy on March 9, 2011 (ADAMS accession number ML110540310). This SDA signified that the NRC staff reviewed the design and found the design met all applicable regulations.

In parallel with the SDA, the NRC staff began preparing a rulemaking to certify the design approved in the SDA. Based on the completion of its safety review, the NRC published a proposed rule on March 24, 2011 (77 FR 16549) that would certify the ESBWR design in Appendix E to 10 CFR Part 52.

In late 2011, while the NRC staff was preparing the final rule, issues were identified with the ESBWR steam dryer, which is a non-safety component. These issues called into question certain conclusions in the staff's safety review under the SDA. Resolution of these issues requires additional analyses by the applicant and review by the NRC staff in order for the NRC staff to conclude the design is acceptable for certification. The design certification rulemaking process is delayed pending resolution of these issues. If the additional analyses resolve the issues, certification, via publication of a final rule, is expected to be completed in 2013.

Section 102 of the National Environmental Policy Act of 1969, as amended (NEPA) (42 USC 4321 *et seq.*), directs that an EIS be prepared for major Federal actions that significantly affect the quality of the human environment. The NRC has implemented Section 102 of NEPA in Title 10 of the Code of Federal Regulations (CFR), Part 51. Further, in 10 CFR 51.20, the NRC has determined that the issuance of a COL under 10 CFR Part 52 is an action that requires an EIS.

The purpose of Detroit Edison's requested NRC action – issuance of the COL – is to obtain a license to construct and operate a new nuclear unit. This license is necessary but not sufficient for construction and operation of the unit. A COL applicant must obtain and maintain the necessary permits from other Federal, State, Tribal, and local agencies and permitting authorities. Therefore, the purpose of the NRC's environmental review of the Detroit Edison application is to determine if a new nuclear power plant of the proposed design can be constructed and operated at the Fermi site without unacceptable adverse impacts on the human environment. The objective of Detroit Edison's anticipated request for USACE action would be to obtain a decision on a permit application proposing structures and/or work in, over, or under navigable waters and/or the discharge of dredged or fill material into waters of the United States, including jurisdictional wetlands. Upon acceptance of the Detroit Edison application, the NRC began the environmental review process described in 10 CFR Part 51 by publishing in the *Federal Register* (FR) a Notice of Intent (73 FR 75142) to prepare an EIS and conduct scoping. On January 14, 2009, the NRC held two scoping meetings in Monroe, Michigan, to obtain public input on the scope of the environmental review. To gather information and to become familiar with the sites and their environs, the NRC and its contractors, Argonne National Laboratory, Energy Research, Inc., and Ecology and Environment, Inc., visited the Fermi site in February 2009 and the four alternative sites, Belle River/St. Clair, Greenwood Energy Center, and two greenfield sites (Petersburg and South Britton sites) in January 2009.

During the Fermi site visit, the NRC staff, its contractors, and the USACE staff met with Detroit Edison staff, public officials, and the public. The NRC staff reviewed the comments received during the scoping process and contacted Federal, State, Tribal, regional, and local agencies to solicit comments. Included in this EIS are (1) the results of the review team's analyses, which consider and weigh the environmental effects of the proposed action (i.e., issuance of the COL) and of building and operating a new nuclear unit at the Fermi site; (2) mitigation measures for reducing or avoiding adverse effects; (3) the environmental impacts of alternatives to the proposed action; and (4) the staff's recommendation regarding the proposed action.

To guide its assessment of the environmental impacts of a proposed action or alternative actions, the NRC has established a standard of significance for impacts based on Council on Environmental Quality guidance (40 CFR 1508.27). Table B-1 of 10 CFR Part 51, Subpart A,

Appendix B, provides the following definitions of the three significance levels – SMALL, MODERATE, and LARGE:

SMALL – Environmental effects are not detectable or are so minor that they will neither destabilize nor noticeably alter any important attribute of the resource.

MODERATE – Environmental effects are sufficient to alter noticeably, but not to destabilize, important attributes of the resource.

LARGE – Environmental effects are clearly noticeable and are sufficient to destabilize important attributes of the resource.

Mitigation measures were considered for each resource category and are discussed in the appropriate sections of the EIS.

In preparing this EIS, the NRC staff and USACE staff reviewed the application, including the Environmental Report (ER) submitted by Detroit Edison; consulted with Federal, State, Tribal, and local agencies; and followed the guidance set forth in NUREG-1555, *Environmental Standard Review Plan*. In addition, the NRC staff considered the public comments related to the environmental review received during the scoping process. Comments within the scope of the environmental review are included in Appendix D of this EIS.

A 75-day comment period began on October 28, 2011, when the U.S. Environmental Protection Agency (EPA) issued a FR Notice of Availability (76 FR 66925) of the draft EIS to allow members of the public to comment on the results of the environmental review. Two public meetings were held on December 15, 2011, at Monroe County Community College, in Monroe, Michigan. During these public meetings, the review team described the results of the NRC environmental review, answered questions related to the review, and provided members of the public with information to assist them in formulating their comments. The comment period for the draft EIS ended January 11, 2012. Comments on the draft EIS and the staff's responses are provided in Appendix E of this EIS.

The USACE issued LRE-2008-00443-1-S11 public notice for a 30-day review on December 23, 2011, describing the proposed USACE-regulated activities associated with the Fermi 3 project; proposed water of the United States avoidance and minimization plan and conceptual mitigation strategy; and USACE preliminary assessment of certain impacts. The purpose of the public notice was to solicit comments from the public; Federal, State, and local agencies and officials; Indian Tribes; and other interested parties in order to consider and evaluate the impacts of regulated activities within the USACE scope of analysis that are associated with the Fermi 3 project. The comments received during the public comment period are under review by USACE.

The NRC staff's recommendation to the Commission related to the environmental aspects of the proposed action is that the COL be issued as requested.[a] This recommendation is based on (1) the application, including the ER submitted by Detroit Edison and the applicant's supplemental letters and responses to the staff's Requests for Additional Information; (2) consultation with other Federal, State, Tribal, and local agencies; (3) the staff's independent review; (4) the staff's consideration of public comments related to the environmental review that were received during the scoping process and on the draft EIS; and (5) the assessments summarized in this EIS, including the potential mitigation measures identified in the ER and this EIS. The USACE will base its evaluation of Detroit Edison's permit application on items (1), (2), (4), and (5) listed above; USACE consideration of public comments received in response to the USACE public notice; the requirements of USACE regulations and the Clean Water Act Section 404(b)(1) Guidelines; and the USACE public interest review. The USACE's permit decision will be based, in part, on this EIS and will be made after issuance of the final EIS and completion of its permit application review and decision-making process.

The NRC staff's evaluation of the site safety and emergency preparedness aspects of the proposed action will be addressed in the NRC's Safety Evaluation Report anticipated to be published in the future.

(a) As directed by the Commission in CLI-12-16, NRC will not issue the COL prior to completion of the ongoing rulemaking to update the Waste Confidence Decision and Rule (see Section 6.1.6 of this EIS).

Abbreviations/Acronyms

χ/Q	dispersion values
°F	degree(s) Fahrenheit
ABWR	advanced boiling water reactor
ac	acre(s)
AC	alternating current
ACHP	Advisory Council on Historic Preservation
ADAMS	Agencywide Documents Access and Management System
ADG	ancillary diesel generator
ADT	average daily traffic
AEC	Atomic Energy Commission
AHS	Auxiliary Heat Sink
ALARA	as low as reasonably achievable
ANSI	American National Standards Institute
APE	area of potential effects
AQCR	Air Quality Control Region
Argonne	Argonne National Laboratory
AST	aboveground storage tank
ASLB	Atomic Safety and Licensing Board
AWEA	American Wind Energy Association
BA	Biological Assessment
BACT	Best Available Control Technology
BEA	Bureau of Economic Analysis (U.S. Department of Commerce)
BEIR	Biological Effects of Ionizing Radiation
BGEPA	Bald and Golden Eagle Protection Act of 1940
BIA	Bureau of Indian Affairs
BiMAC	basemat internal melt arrest and coolability
BMP	best management practice
Bq	Becquerel
Bq/MTU	Becquerel per metric ton uranium
BRC	Blue Ribbon Commission
Btu	British thermal unit(s)
BWR	boiling water reactor
CAA	Clean Air Act
CAES	compressed air energy storage
CAIR	Clean Air Interstate Rule

CCR	coal combustion residuals
CCRG	Commonwealth Cultural Resources Group, Inc.
CCS	carbon capture and sequestering/sequestration
CDC	Centers for Disease Control and Prevention
CDF	core damage frequency
CEQ	Council on Environmental Quality
CER	Capital Expenditure and Recovery
CFR	Code of Federal Regulations
cfs	cubic feet per second
cfu	colony forming units
CH_4	methane
CHP	combined heat and power
Ci	curie(s)
CIRC	Circulating Water System
CIS	containment isolation system
CN	Canadian National
CNF	Capacity Need Forum (MPSC)
CO	carbon monoxide
CO_2	carbon dioxide
CO_2-e	carbon dioxide-equivalent
COL	combined construction permit and operating license
CSAPR	Cross-State Air Pollution Rate
CSP	concentrated solar power
CSX	CSX Transportation
CT	combustion turbine
CWA	Clean Water Act
CWIS	Cooling Water Intake Structure
CZMA	Coastal Zone Management Act
DA	Department of the Army
dB	decibel
dBA	A-weighted decibel
DBA	design-basis accident
dbh	diameter at breast height
DC	direct current
DCD	Design Control Document
DDT	dichlorodiphenyltrichloroethane
Detroit Edison	Detroit Edison Company
DHS	U.S. Department of Homeland Security
DNL	equivalent continuous sound level

DNR	Designated Network Resource
DOC	U.S. Department of Commerce
DOD	U.S. Department of Defense
DOE	U.S. Department of Energy
DOI	U.S. Department of the Interior
DOT	Department of Transportation
D/Q	deposition factor
DRIWR	Detroit River International Wildlife Refuge
DSM	demand-side management
DTW	Detroit Metropolitan Wayne County Airport
DWSD	Detroit Water and Sewerage Department
E&E	Ecology and Environment, Inc.
EAB	Exclusion Area Boundary
EERE	U.S. Department of Energy Office of Energy Efficiency and Renewable Energy
EGS	engineered geothermal system
EIA	Energy Information Administration
EIS	environmental impact statement
ELF	extremely low frequency
EMF	electromagnetic field
EOP	emergency operating procedure
EPA	U.S. Environmental Protection Agency
EPRI	Electric Power Research Institute
EPT	Ephemeroptera, Plecoptera, Trichoptera (index)
EPZ	emergency planning zone
ER	Environmental Report
ERI	Energy Research, Inc.
ESA	Endangered Species Act of 1973, as amended
ESBWR	Economic Simplified Boiling Water Reactor
ESRP	Environmental Standard Review Plan
FAA	Federal Aviation Administration
FEMA	Federal Emergency Management Agency
FERC	Federal Energy Regulatory Commission
Fermi	Enrico Fermi Atomic Power Plant
Fermi 1	Enrico Fermi Unit 1
Fermi 2	Enrico Fermi Unit 2
Fermi 3	Enrico Fermi Unit 3
FES	Final Environmental Statement
FIRM	Flood Insurance Rate Map
FIS	Financial Reporting and Analysis

FP	fire pump
fps	feet per second
FPS	Fire Protection System
FR	*Federal Register*
FSAR	Final Safety Analysis Report
FSER	Final Safety Evaluation Report
ft	foot (feet)
ft/day	feet per day
ft^3	cubic feet
FTE	full-time equivalent
FWS	U.S. Fish and Wildlife Service
FY	fiscal year
GAF	Generation and Fuel
gal	gallon
GBq	gigabecquerel
GC	gas centrifuge
GD	gaseous diffusion
GEH	General Electric-Hitachi Nuclear Energy Americas, LLC
GEIS	*Generic Environmental Impact Statement for License Renewal of Nuclear Plants*
GEIS-DECOM	*Generic Environmental Impact Statement for Decommissioning of Nuclear Facilities: Supplement 1, Regarding the Decommissioning of Nuclear Power Reactors*
GHG	greenhouse gas
GIS	geographical information system
GLC	Great Lakes Commission
GLENDA	Great Lakes Environmental Database
GLOFS	Great Lakes Operational Forecast System
GLWC	Great Lakes Wind Council
gpd	gallon(s) per day
gpm	gallon(s) per minute
GWh	gigawatt hour(s)
GWP	global warming potential
ha	hectare
HAP	hazardous air pollutant
HCMA	Huron-Clinton Metropolitan Authority
HDR	hot dry rock
HEPA	high-efficiency particulate air
HFC	hydrofluorocarbon

HFE	hydrofluorinated ether
HLW	high-level waste
HQUSACE	U.S. Army Corps of Engineers Headquarters
hr	hour(s)
HRSG	heat recovery steam generator
HUD	U.S. Department of Housing and Urban Development
HVAC	heating, ventilating, and air-conditioning
IAEA	International Atomic Energy Agency
ICRP	International Commission on Radiological Protection
IEEE	Institute of Electrical and Electronics Engineers
IGCC	integrated gasification combined cycle
IGLD 85	International Great Lakes Datum of 1985
IJC	International Joint Commission
in.	inch(es)
INAC	Indian and Northern Affairs Canada
IOU	investor-owned utility
IPCC	Intergovernmantal Panel on Climate Change
IPCS	Integrated Plant Computer System
IPP	independent power producer
IRP	Integrated Resource Plan
ISD	Intermediate School District
ISFSI	Independent Spent Fuel Storage Installation
ITC	ITC Holdings Corporation
JPA	Joint Permit Application
kg	kilogram(s)
KiKK	Childhood Cancer in the Vicinity of Nuclear Power Plants (German acronym)
km	kilometer(s)
km^2	square kilometer(s)
kV	kilovolt(s)
kW	kilowatt(s)
kWh	kilowatt hour(s)
L	liter(s)
L_{90}	sound level exceeded 90 percent of the time
LaMP	Lakewide Management Plan
lb	pound(s)
L_{dn}	day-night average sound level
LEDPA	least environmentally damaging practicable alternative

LEOFS	Lake Erie Operational Forecast System
L_{eq}	equivalent continuous sound level
LET	Lake Erie Transit
LFA	Load Forecasting Adjustment
LLW	low-level waste
LOLE	Loss of Load Expectation
LOLP	Loss-of-Load Probability
LOS	level of service
LPZ	low population zone
LRF	large release frequency
LTRA	Long-Term Reliability Assessment (NERC)
LW	long wave
LWR	light water reactor
µg	microgram(s)
m	meter(s)
m^3	cubic meter(s)
MACCS2	MELCOR Accident Consequence Code System
MBTA	Migratory Bird Treaty Act of 1918
MCCC	Monroe County Community College
mCi	millicurie
MCL	maximum contaminant level; Michigan Compiled Laws
MCRC	Monroe County Road Commission
MDCH	Michigan Department of Community Health
MDCT	mechanical draft cooling tower
MDELEG	Michigan Department of Energy, Labor and Economic Growth
MDEQ	Michigan Department of Environmental Quality
MDNR	Michigan Department of Natural Resources
MDOT	Michigan Department of Transportation
MDSP	Michigan Department of State Police
MEI	maximally exposed individual
METC	Michigan Electric Transmission Company
mGy	milliGray
MGD	million gallons per day
mi	mile(s)
mi^2	square mile(s)
MichCon	Michigan Consolidated Gas Company
MISO	Midwest Independent System Operator
MIT	Massachusetts Institute of Technology
mL	milliliter(s)
MMT	million metric tons

MMTCO$_2$-e	million metric tons of carbon dioxide equivalent
MNFI	Michigan Natural Features Inventory
mo	month(s)
MOA	Memorandum of Agreement
MOU	Memorandum of Understanding
mph	mile(s) per hour
MPSC	Michigan Public Service Commission
mrad	milliradian
mrem	millirem(s)
MSA	Metropolitan Statistical Area
MSW	municipal solid waste
MT	metric ton(s) (or tonne[s])
MTEP	MISO Transmission Expansion Plan
MTU	metric ton(s) of uranium
MW	megawatt(s)
MW(e)	megawatt(s) electrical
MW(t)	megawatt(s) thermal
MWd	megawatt-day(s)
MWd/MTU	megawatt-day(s) per metric ton of uranium
MWh	megawatt hour(s)
NAAQS	National Ambient Air Quality Standard
NACD	Native American Consultation Database
NaCl	sodium chloride
NAGPRA	Native American Graves Protection and Repatriation Act of 1990
NAS	National Academy of Sciences
NAVD 88	North American Vertical Datum of 1988
DCDC	National Climate Data Center
NCI	National Cancer Institute
NCRP	National Council on Radiation Protection and Measurements
NDCT	natural draft cooling tower
NEI	Nuclear Energy Institute
NEPA	National Environmental Policy Act of 1969, as amended
NERC	North American Electric Reliability Corporation
NESC	National Electrical Safety Code
NESHAP	National Emission Standards for Hazardous Air Pollutants
NF$_3$	nitrogen trifluoride
NGCC	natural gas combined-cycle
NHPA	National Historic Preservation Act of 1966, as amended
NIEHS	National Institute of Environmental Health Sciences
NMFS	National Marine Fisheries Service

NML	noise monitoring location
NNW	north-northwest
N_2O	nitrous oxide
NO_2	nitrogen dioxide
NOAA	National Oceanic and Atmospheric Administration
NO_x	nitrogen oxide
NPDES	National Pollutant Discharge Elimination System
NPHS	normal power heat sink
NPS	National Park Service
NRC	U.S. Nuclear Regulatory Commission
NRCS	Natural Resources Conservation Service
NREL	National Renewable Energy Laboratory
NREPA	Natural Resources and Environmental Protection Act
NRHP	*National Register of Historic Places*
NS	Norfolk Southern
NSPS	New Source Performance Standard
NSR	new source review
NTC	Nuclear Training Center
NTU	nephelometric turbidity unit
NWI	National Wetland Inventory
NWIS	National Water Information System
NWR	National Wildlife Refuge
O_3	ozone
ODCM	Offsite Dose Calculation Manual
ODNR	Ohio Department of Natural Resources
OGS	off-gas system
OSHA	Occupational Safety and Health Administration
PAM	primary amebic meningoencephalitis
PAP	personnel access portal
Pb	lead
PC	personal computer
PCB	polychlorinated biphenyl
pCi/L	picocurie(s) per liter
PCTMS	Plant Cooling Tower Makeup System
PEM	palustrine emergent marsh
PESP	Pesticide Environmental Stewardship Program
PFC	perfluorocarbon
PFO	palustrine forested wetland
P-IBI	Planktonic Index of Biotic Integrity

PIPP	Pollution Incident Prevention Plan
PJM	PJM Interconnection
PM	particulate matter
$PM_{2.5}$	particulate matter with a mean aerodynamic diameter of less than or equal to 2.5 μm
PM_{10}	particulate matter with a mean aerodynamic diameter of less than or equal to 10 μm
PRA	probabilistic risk assessment
PRB	Powder River Basin
PSD	Prevention of Significant Deterioration
psia	pounds per square inch absolute
PSR	Physicians for Social Responsibility
PSS	palustrine scrub-shrub wetland
PSWS	Plant Service Water System
PTE	potential to emit
Pu-239	plutonium-239
PV	photovoltaic
PWSS	pretreated water supply system
RAI	Request for Additional Information
RCRA	Resource Conservation and Recovery Act of 1976, as amended
RDF	refuse-derived fuel
REIRS	Radiation Exposure Information and Reporting System
rem	roentgen equivalent man
REMP	radiological environmental monitoring program
RESA	Regional Educational Service Agency
RFC	Reliability*First* Corporation
RHAA	Rivers and Harbors Appropriation Act of 1899
RHR	residual heat removal
RIMS II	Regional Input-Output Modeling System
ROI	region of interest
ROW	right-of-way
RPS	Renewable Portfolio Standard
RRD	Remediation and Redevelopment Division
RSICC	Radiation Safety Information Computational Center
RTO	Regional Transmission Organization
RTP	Regional Transportation Plan
RV	recreational vehicle
Ryr	reactor-year

SACTI	Seasonal/Annual Cooling Tower Impact
SAMA	severe accident mitigation alternative
SAMDA	severe accident mitigation design alternative
SAMG	severe accident management guidelines
SBO	station blackout
SCPC	supercritical pulverized coal
SCR	selective catalytic reduction
SDA	standard design approval
SDG	standby diesel generator
sec	second(s)
SEGS	Solar Energy Generating System
SEMCOG	Southeast Michigan Council of Governments
SER	Safety Evaluation Report
SESC	soil erosion and sedimentation control
SF_6	sulfur hexafluoride
SHPO	State Historic Preservation Office(r)
SO_2	sulfur dioxide
SO_x	sulfur oxides
SOARCA	State-of-the-Art Reactor Consequence Analyses
SRHP	*State Register of Historic Places*
SRREN	Special Report on Renewable Energy Sources and Climate Change Mitigation
SSC	system, structure, and component
SSE	safe shutdown earthquake ground motion
STG	steam turbine generator
STORET	Storage and Retrieval Database
SUV	sport-utility vehicle
Sv	sievert
SWMS	solid radioactive waste management system
SWPPP	Stormwater Pollution Prevention Plan
SWS	Station Water System
TDS	total dissolved solids
TEDE	total effective dose equivalent
THPO	Tribal Historic Preservation Office
TI	Temporary Instruction
TIP	Transportation Improvement program
TLD	thermoluminescent dosimeter
TMDL	total maximum daily load
TRAGIS	Transportation Routing Analysis Geographic Information System
TRU	transuranic

U.S.	United States
USC	United States Code
U_3O_8	triuranium octoxide ("yellowcake")
UF_6	uranium hexafluoride
UMTRI	University of Michigan Transportation Research Institute
UO_2	uranium dioxide
USACE	U.S. Army Corps of Engineers
USBLS	U.S. Bureau of Labor Statistics
USCB	U.S. Census Bureau
USDA	U.S. Department of Agriculture
USGCRP	U.S. Global Change Research Program
USGS	U.S. Geological Survey
VIB	Vehicle Inspection Building
VOC	volatile organic compound
WHO	World Health Organization
WNW	west-northwest
WPSCI	Wolverine Power Supply Cooperative, Inc.
WRA	Wind Resource Area
WTE	waste-to-energy
WWSL	wastewater stabilization lagoon
WWTP	wastewater treatment plant
yd^3	cubic yard(s)
yr	year(s)

Appendix F

Key Consultation Correspondence

Appendix F

Key Consultation Correspondence

This appendix identifies the consultation correspondence sent and received during the environmental review of the Enrico Fermi Unit 3 (Fermi 3) combined license application. Table F-1 presents correspondence related to historic properties and cultural resource, and Table F-2 presents correspondence related to natural resources. In addition, a copy of the Biological Assessment (BA) and consultation correspondence with the U.S. Fish and Wildlife Service concerning the BA and the signed Memorandum of Agreement (MOA) between the U.S. Nuclear Regulatory Commission and the Michigan State Historic Preservation Officer Regarding the Demolition of the Enrico Fermi Atomic Power Plant, Unit 1, located in Monroe County, Michigan, are included in this appendix.

Table F-1. List of Consultation Correspondence Related to Historic Properties and Cultural Resources

Source	Recipient	Date Accession No.
U.S. Nuclear Regulatory Commission (Gregory P. Hatchett)	Advisory Council on Historic Preservation (Don Klima)	December 24, 2008 ML083151399
U.S. Nuclear Regulatory Commission (Gregory P. Hatchett)	Keweenaw Bay Indian Community (Warren C. Swartz)	December 24, 2008 ML083190398
U.S. Nuclear Regulatory Commission (Gregory P. Hatchett)	Bay Mills Indian Community (Jeffery D. Parker)	December 24, 2008 ML083190083
U.S. Nuclear Regulatory Commission (Gregory P. Hatchett)	Grand Traverse Band of Ottawa and Chippewa Indians (Robert Kewaygoshkum)	December 24, 2008 ML083190375
U.S. Nuclear Regulatory Commission (Gregory P. Hatchett)	Lac Vieux Desert Band of Lake Superior Chippewa Indians (James Williams, Jr.)	December 24, 2008 ML083190406
U.S. Nuclear Regulatory Commission (Gregory P. Hatchett)	Little Traverse Bay Bands of Odawa Indians (Frank Ettawageshik)	December 24, 2008 ML083190425
U.S. Nuclear Regulatory Commission (Gregory P. Hatchett)	Pokagon Band of Potawatomi Indians (John A. Miller)	December 24, 2008 ML083190442

Table F-1. (contd)

Source	Recipient	Date Accession No.
U.S. Nuclear Regulatory Commission (Gregory P. Hatchett)	Sault Ste. Marie Tribe of Chippewa Indians of Michigan (Aaron Payment)	December 24, 2008 ML083190489
U.S. Nuclear Regulatory Commission (Gregory P. Hatchett)	Hannahville Indian Community (Kenneth Meshigaud)	December 24, 2008 ML083190379
U.S. Nuclear Regulatory Commission (Gregory P. Hatchett)	Huron Potawatomi, Inc. (Laura Spurr)	December 24, 2008 ML083190382
U.S. Nuclear Regulatory Commission (Gregory P. Hatchett)	Saginaw Chippewa Indian Tribe of Michigan (Fred Cantu, Jr.)	December 24, 2008 ML083190448
U.S. Nuclear Regulatory Commission (Gregory P. Hatchett)	Match-e-be-nash-she-wish Band of Pottawatomi Indians of Michigan (David K. Sprague)	December 24, 2008 ML083190436
U.S. Nuclear Regulatory Commission (Gregory P. Hatchett)	Little River Band of Ottawa Indians (Larry Romanelli)	December 24, 2008 ML083190415
U.S. Nuclear Regulatory Commission (Gregory P. Hatchett)	Michigan State Historic Preservation Officer (Brian D. Conway)	December 24, 2008 ML083151405
U.S. Nuclear Regulatory Commission (Gregory P. Hatchett)	Forest County Potawatomi (Harold G. Frank)	December 31, 2008 ML083520641
U.S. Nuclear Regulatory Commission (Gregory P. Hatchett)	Shawnee Tribe (Ron Sparkman)	December 31, 2008 ML083530066
U.S. Nuclear Regulatory Commission (Gregory P. Hatchett)	Delaware Nation (Edgar L. French)	December 31, 2008 ML083530050
U.S. Nuclear Regulatory Commission (Gregory P. Hatchett)	Wyandotte Nation (Leaford Bearskin)	December 31, 2008 ML083530077
U.S. Nuclear Regulatory Commission (Gregory P. Hatchett)	Ottawa Tribe of Oklahoma (Charles Todd)	December 31, 2008 ML083530043
U.S. Nuclear Regulatory Commission (Bruce A. Watson)	Michigan State Historic Preservation Officer (Brian D. Conway)	December 2, 2010 ML101790096
U.S. Nuclear Regulatory Commission (Ryan Whited)	Michigan State Historic Preservation Officer (Brian D. Conway)	December 16, 2010 ML101820302
U.S. Nuclear Regulatory Commission (Brent Clayton)	Advisory Council on Historic Preservation (Reid Nelson)	October 13, 2011 ML112500143

Table F-1. (contd)

Source	Recipient	Date Accession No.
Advisory Council on Historic Preservation (LaShavio Johnson)	U.S. Nuclear Regulatory Commission (Brent Clayton)	October 25, 2011 ML112990031
U.S. Nuclear Regulatory Commission (John Fringer)	Michigan State Historic Preservation Officer (Martha M. Faes)	August 22, 2011 ML112070027
U.S. Nuclear Regulatory Commission (John Fringer)	Michigan State Historic Preservation Officer (Martha M. Faes)	August 24, 2011 ML112070043
U.S. Nuclear Regulatory Commission (John Fringer)	Interested party (Donald Ferencz)	November 17, 2011 ML12129A340
U.S. Nuclear Regulatory Commission (John Fringer)	Interested party (Philip Harrigan)	November 17, 2011 ML12129A348
U.S. Nuclear Regulatory Commission (John Fringer)	Interested party (David Nixon)	November 17, 2011 ML12129A343
U.S. Nuclear Regulatory Commission (John Fringer)	Interested party (Christine Kull)	November 17, 2011 ML12129A350
U.S. Nuclear Regulatory Commission (John Fringer)	Interested party (Mike Hartman)	November 17, 2011 ML12129A339
U.S. Nuclear Regulatory Commission (John Fringer)	American Nuclear Society (Laura Scheele)	November 17, 2011 ML12129A341
U.S. Nuclear Regulatory Commission (John Fringer)	Interested party (James Walther)	November 17, 2011 ML12129A345
Interested party (Donald Ferencz)	U.S. Nuclear Regulatory Commission (John Fringer)	November 17, 2011 ML12129A355
Interested party (David Nixon)	U.S. Nuclear Regulatory Commission (John Fringer)	November 17, 2011 ML12129A344
Interested party (Christine Kull)	U.S. Nuclear Regulatory Commission (John Fringer)	November 18, 2011 ML12129A359
Interested party (James Walther)	U.S. Nuclear Regulatory Commission (John Fringer)	November 21, 2011 ML12129A361
Interested party (Philip Harrigan)	U.S. Nuclear Regulatory Commission (John Fringer)	December 1, 2011 ML12129A360
American Nuclear Society (Laura Scheele)	U.S. Nuclear Regulatory Commission (John Fringer)	December 19, 2011 ML12143A465
U.S. Nuclear Regulatory Commission (Scott C. Flanders)	Advisory Council on Historic Preservation (Reid Nelson)	March 7, 2012 ML120450110

Table F-2. List of Consultation Correspondence Related to Natural Resources

Source	Recipient	Date Accession No.
U.S. Nuclear Regulatory Commission (Gregory P. Hatchett)	U.S. Fish and Wildlife Service (Craig Czarnecki)	December 23, 2008 ML083151398
U.S. Nuclear Regulatory Commission (Gregory P. Hatchett)	National Marine Fisheries Service (Mary Colligan)	December 24, 2008 ML083151403
U.S. Nuclear Regulatory Commission (Gregory P. Hatchett)	Ohio Department of Natural Resources (Patricia Jones)	December 24, 2008 ML083151404
U.S. Nuclear Regulatory Commission (Gregory P. Hatchett)	Great Lakes Fisheries Commission (Kelley Smith)	December 24, 2008 ML083151400
U.S. Nuclear Regulatory Commission (Gregory P. Hatchett)	International Joint Commission (James G. Chandler)	December 24, 2008 ML083151401
U.S. Nuclear Regulatory Commission (Gregory P. Hatchett)	Michigan Natural Features Inventory (Leni Wilsmann)	December 24, 2008 ML083151402
U.S. Nuclear Regulatory Commission (Gregory P. Hatchett)	Michigan Department of Environmental Quality (Steven Chester)	December 31, 2008 ML083590138
National Marine Fisheries Service (Mary A. Colligan)	U.S. Nuclear Regulatory Commission (Gregory P. Hatchett)	January 21, 2009 ML090711069
U.S. Fish and Wildlife Service (Craig Czarnecki)	U.S. Nuclear Regulatory Commission (Gregory P. Hatchett)	January 28, 2009 ML090750973
Michigan Department of Environmental Quality (Elizabeth M. Browne)	U.S. Nuclear Regulatory Commission	February 3, 2009 ML0906504561
Michigan Department of Natural Resources (Lori Sargent)	U.S. Nuclear Regulatory Commission (Gregory P. Hatchett)	February 9, 2009 ML090401015
National Marine Fisheries Service (Mary A. Colligan)	U.S. Nuclear Regulatory Commission (Ryan Whited)	November 17, 2011 ML11336A064
U.S. Department of the Interior (Lisa Chetnik Treichel)	U.S. Nuclear Regulatory Commission (Bruce Olson)	January 9, 2012 ML12026A464
U.S. Nuclear Regulatory Commission (Anthony H. Hsia)	U.S. Fish and Wildlife Service (Scott Hicks)	March 30, 2012 ML120260586
U.S. Fish and Wildlife Service (Scott Hicks)	U.S. Nuclear Regulatory Commission (Anthony H. Hsia)	June 8, 2012 ML12178A137

MEMORANDUM OF AGREEMENT
BETWEEN
THE U.S. NUCLEAR REGULATORY COMMISSION
AND THE MICHIGAN STATE HISTORIC PRESERVATION OFFICER
REGARDING THE DEMOLITION OF THE
ENRICO FERMI ATOMIC POWER PLANT, UNIT 1 FACILITY LOCATED
IN MONROE COUNTY, MICHIGAN
SUBMITTED TO THE ADVISORY COUNCIL ON HISTORIC PRESERVATION
PURSUANT TO 36 CFR 800.6(b)(1)

WHEREAS, the United States Nuclear Regulatory Commission (NRC), through its review of the Fermi Atomic Power Plant, Unit 3 (Fermi 3) combined license (COL) application pursuant to 10 CFR 51, has determined that the construction of the proposed Fermi 3 facility will have an adverse effect upon the Enrico Fermi Atomic Power Plant (Fermi 1), which appears to meet the criteria for listing in the National Register of Historic Places (NRHP); and

WHEREAS, the NRC has consulted with the Michigan State Historic Preservation Officer (SHPO) pursuant to 36 CFR Part 800, regulations implementing Section 106 of the National Historic Preservation Act (16 USC 470f); and

WHEREAS, the NRC has invited Detroit Edison Company (DTE), as owner of the Fermi 1 property and NRC general licensee pursuant to 10 CFR Part 50, to be a signatory to this Memorandum of Agreement (MOA) in accordance with 36 CFR 800.6(c)(2);

NOW, THEREFORE, the NRC, DTE, and the SHPO agree that the demolition of Fermi 1 (Project) shall be implemented in accordance with the following stipulations in order to take into account the effects of the Project on historic properties.

STIPULATIONS

DTE shall notify the NRC and the Michigan SHPO of completion of Stipulations I and II prior to the demolition of the Fermi 1 structure.

    I.   RECORDATION

        A.   DTE will document Fermi I so that there is a permanent record of its existence. The recordation packages shall follow the SHPO *Documentation Guidelines* (Appendix A) and shall be submitted to the SHPO for review and approval.

        B.   The completed Fermi 1 documentation package shall be submitted to the SHPO for review within one (1) year of the date of this agreement. The approved original documentation package shall be submitted to the SHPO for deposit in the State Archives of Michigan and another original copy of the documentation shall be submitted to the Monroe County Library and Reference Center.

Fermi 1 MOA   p. 1 of 3   March 8, 2012

II. EXHIBIT

DTE, in consultation with Monroe County Community College and other interested parties and the SHPO, shall develop and establish a permanent public exhibit regarding the history of the Fermi 1 Plant within 2 years of the execution of this agreement. DTE will coordinate with the parties to develop a mutually acceptable plan for the scope, location, and design of this exhibit. At the completion (i.e., conclusion) of the exhibit, DTE shall offer any remaining archival items pertaining to the history of Fermi 1 to local, State and Federal agencies and non-profit organizations potentially interested in permanent retention or display of these items.

III. AMENDMENT AND DURATION

The NRC, the SHPO or DTE may propose to the other parties that this MOA be amended, whereupon the parties will consult in accordance with 36 CFR 800.6(c)(7) to consider such an amendment.

If the terms of this MOA have not been implemented within three (3) years of its execution, this MOA shall be considered null and void. In such event, DTE shall so notify the parties to this MOA, and if NRC chooses to continue with the undertaking, shall re-initiate review of the undertaking in accordance with 36 CFR 800.

IV. DISPUTE RESOLUTION

Disputes regarding the completion of the terms of this agreement shall be resolved by consultation between the signatories. If, within thirty (30) days of an objection to this agreement, the signatories cannot agree on a resolution, any one of the signatories may request the participation of the Advisory Council on Historic Preservation (Council) to assist in resolving the dispute.

V. TERMINATION

Upon completion of Stipulations I and II, if this MOA is not amended following the consultation set out in Stipulations III and IV, it may be terminated by any signatory or invited signatory. The signatory proposing to terminate this MOA shall so notify the other signatories, explaining the reasons for termination and affording them at least 30 days to consult and seek alternatives to termination. Within 30 days following this notification of termination, any one of the signatories shall notify the other signatories if it will: a) initiate consultation to execute a subsequent MOA that explicitly terminates or supersedes its terms; or b) request the comments of the Council under 36 CFR 800.7(a) and proceed accordingly.

Execution of this MOA by the NRC, DTE, and the Michigan SHPO and implementation of its terms evidence the NRC has afforded the Council an opportunity to comment on the Project and its effects on historic properties and the NRC has taken into account the effects of the Project on historic properties.

SIGNATORIES:

UNITED STATES NUCLEAR REGULATORY COMMISSION

By: _____ Date: 3/7/2012

     Scott Flanders, Director,
     Office of New Reactors, Division of Site Safety and Environmental Analysis

MICHIGAN STATE HISTORIC PRESERVATION OFFICER

By: _____ Date: 3/22/12

     Brian D. Conway, State Historic Preservation Officer

INVITED SIGNATORIES:

DETROIT EDISON COMPANY

By: _____ Date: 3/14/2012

     Peter W. Smith, Director
     Nuclear Development - Licensing & Engineering

MONROE COUNTY COMMUNITY COLLEGE

By: _____ Date: 3/19/12

     Dr. David Nixon, President

Appendix A: MICHIGAN STATE HISTORIC PRESERVATION OFFICE DOCUMENTATION
              GUIDELINES

Fermi 1 MOA   p. 3 of 3   March 8, 2012

Appendix F

MICHIGAN STATE HISTORIC PRESERVATION OFFICE
DOCUMENTATION GUIDELINES

The following guidelines provide instruction for producing permanent documentation of historic properties Following submittal to the State Historic Preservation Office. the photos produced will be transferred to the State Archives. where they will be maintained and made available to the public for research purposes. In many cases, this documentation will constitute the only visual public record of a resource. It is therefore important that reports, drawings and photographs adequately depict the salient visual characteristics of the resource, and that they be produced using archivally stable materials and procedures.

The specifications outlined in this memorandum are intended to ensure that the material will be of high quality and remain in usable condition for many years to come. The guidelines were adapted from those used for submitting nominations to the National Register of Historic Places, as described in **National Register Bulletin 16:** *Guidelines for Completing National Register of Historic Places Forms.* The complete text of this and other National Register Bulletins may be found on the web at *http://www.nps.gov/history/nr/publications/*.

I. REPORTS - GENERAL INSTRUCTIONS

Reports should be printed on archival paper and be 8½ by 11 inches in size.

II. DESCRIPTIVE AND HISTORICAL NARRATIVES

The report should contain a descriptive and historical narrative about the resource(s). The descriptive overview should concisely but thoroughly describe the resource. including discussion of its site and setting: overall design and form. dimensions. structural character, materials, decorative or other details. and alterations. The historical narrative should provide an account of the resource's history and explain its significance in terms of the national register criteria (information about the criteria for listing a resource in the national register may be found on the web at http://www.nps.gov/history/nr/publications/bulletins/nrb15/nrb15_2.htm). Published and unpublished sources should be used as needed to document the resource's significance. For bridges and public structures, public records and newspapers should be used for information concerning the historical background and construction of the resource and to identify those involved in its design and construction. All sources of information (including author. title. publisher, date of publication, volume and page number) should be listed in a bibliography.

III. MAPS

Documentation for the historical narrative must include one or more maps that encompass the whole development, including:

- **USGS Map** – an original United States Geological Survey (USGS) topographical map indicating the location of the subdivision and listing its UTM coordinates.

- **Other Map(s)** - The maps must show the locations of all historic and non-historic features of districts and complexes. If more than one map is required to cover the entire district, a key map should illustrate the entire district and its boundaries.

Information District Maps Must Provide

- District or property name

- Name of community, county, and state

- Significant natural features such as lakes and rivers, with names

- All streets, railroad lines, old railroad grades, and any other transportation rights of way, labeled in bold print with their names

- Lot or property lines

- Outlines or representations for all surveyed properties

- Patterned coding of footprints or representations of all buildings to indicate whether they are contributing or non-contributing to the district's or complex's historic character and significance. The outlines or representations of contributing resources must be darkened, while they are left light for non-contributing resources.

- For districts, street addresses for all properties listed in the description's inventory section; if the properties have numbered street addresses, no other form of identification may appear on the map.

- Boundary of the property associated with the district or complex property.

- Key identifying any symbols used

- North directional arrow

- Scale bar (in case map is copied in larger or smaller format)

Do Not:

- Use color coding. Photocopying in black and white will render color coding unreadable.

Map Standards

The final copies of maps must be printed on white paper meeting the national register's standards for archival stability – 20 pound acid-free paper with a two percent alkaline reserve. Two **original** copies must be provided of all maps and site plans. Tape, staples, and adhesive labels may not be used. Maps should be in 8 ½" X 11" format, if possible. Map sheets larger than 11" X 17" are not acceptable.

The district map should show both the lot lines and the outlines of the buildings. For business districts containing buildings that occupy most of their lots, the maps must show the building outlines. Outside of business districts, surveyed buildings can be shown by square boxes if maps showing building outlines are not available. Monuments and other objects may be represented by circles or dots.

Page 2 of 6

IV.   DRAWINGS - GENERAL INSTRUCTIONS

Drawings should be drawn or printed on archival paper and folded to fit an archival folder approximately 8½ by 11 inches. Use coding, crosshatching, numbering, transparent overlays, or other standard graphic techniques to' indicate the information. Do not use color because it can not be reproduced by microfilming or photocopying. Drawings should be used to document the existing condition of the resource, the evolution of a resource, alterations to a building or complex of buildings, floor plans of interior spaces. - Site plans should have a graphic north arrow and include locations and types of trees, shrubs and planting beds. All architectural and site plans should include dimensions indicating the overall size of buildings, sizes of major interior spaces and distances between major site features. If original drawings of the resource(s) exist, add a graphic scale the drawings and reproduce them to fit on 8½ by 11 inch archival paper. Photographic reductions are permissible provided they meet the photographic requirements specified in these guidelines.

V.   PHOTOGRAPHS - GENERAL INSTRUCTIONS

Submit clear and descriptive photographs and negatives in acid-free envelopes. Photographs should provide a clear visual representation of the historic integrity and significant features of the resource. The number of photographs needed will vary according to the project and the nature of the resource. The attached article by David Ames, A *Primer on Architectural Photography and the Photo Documentation of Historic Structures* (Vernacular Architecture Forum News, no date) provides helpful information for photographing buildings and structures. This article is available on the web at *http://dspace.udel.edu:8080/dspace/bitstream/19716/2831/1/A%20primer%20on.pdf.*

GUIDELINES FOR PHOTOGRAPHIC COVERAGE

Photography should include at least two general views of each building to be demolished, each if possible showing two sides, so that all four sides are photographed, plus at least one streetscape view looking in each direction of the part of the street in which each building is located. Thus, for each building, six views, unless several buildings are in one short stretch of the same street. If there are any examples left of any of the same building form that retain a high state of integrity, photos should be taken of one sample building for each building form, two views of each together showing all four sides.

Buildings, Structures and Objects

- Submit one or more views to show the principal facades and the environment or setting in which the resource is located;

- Additions, alterations, intrusions, and dependencies should appear in the photographs;

- Include views of interiors, outbuildings, landscaping, or unusual details if the significance of the resource is entirely or in part based on them.

Page 3 of 6

Historic and Archaeological Sites

- Submit one or more photographs to depict the condition of the site and any aboveground or surface features and disturbances;

- If they are relevant to the site's significance, include drawings or photographs that illustrate artifacts that have been removed from the site;

- At least one photograph should show the physical environment and configuration of the land making up the site.

BASIC TECHNICAL REQUIREMENTS

Photographs must be:

- at least 5 x 7 inches, preferably 8 x 10 inches, unmounted (do not affix the photographs to paper, cards, or any other material); photographs with borders are preferred;

- submitted in acid free envelopes; the envelopes should be labeled in pencil (see labeling instructions below).

Envelope Labeling Instructions

Neatly print the following information on the upper right corner of the envelope in soft lead pencil:

1. Name of the resource;
2. Street Address, township, county, and state where the resource is located;
3. Name of photographer;
4. Date of photograph;
5. Description of view indicating direction of camera;
6. Photograph number.

Do not use adhesive labels for this information.

Film Photography

- Photographs must be printed on double or medium-weight black-and-white paper having a matte, glossy, or satin finish; fiber-based papers are preferred; resin-coated papers that have been processed automatically will be accepted provided they have been properly processed and thoroughly washed; we recommend the use of a hypo-clearing or neutralizing agent, and toning in selenium or sepia to extend the useful life of the photographs;

- The negatives must be submitted with the prints. Each strip of negatives should be submitted in acid free envelopes that have the following information submitted in soft lead pencil in the upper right corner of the envelope.

1. Name of the resource;
2. Name of the photographer;

Page 4 of 6

3. Date of photograph;
4. Negative numbers

Digital Photography

Camera:

BEST: At Least 6 megapixel digital SLR Camera
Acceptable: Minimum 6 megapixel point-and-shoot digital camera
Acceptable: 2 – 5 megapixel SLR or point-and-shoot digital camera
Not acceptable:

- Camera phones
- Disposable or single-use digital cameras
- Digital cameras with fewer than 2 megapixels of resolution

Image format:

BEST: First generation Tag image file format (TIFF) or RAW
Acceptable:

- Joint Photographic Experts Group (JPEG) converted to TIFF
- JPEG must not be altered in any way prior to conversion
- After the image has been saved as a TIFF, use the guidelines outlined in the section titled "Labeling the Image.

Capturing the Image:

BEST: Minimum 6 megapixels (2000 x 3000 pixel image) at 300 dpi
Acceptable: Minimum 2 megapixels (1200 x 1600 pixel image) at 300 dpi

Printer paper and inks[1]:

BEST Inks: Manufacturer recommended pigmented ink for photograph printing
- Some examples:
  - Epson UltraChrome K3
  - Kodak No. 10 Pigmented Inks
  - HP Vivera Pigment Inks
  - Epson Claria "Hi-Definition Inks"
  - Epson DuraBrite Ultra Pigmented Inks
  - HP Vivera 95 dye-based inks

BEST Papers: Photographic Matte Paper
Not acceptable:
- Regular copy or printer papers
- Glossy photographic paper papers
- Paper or ink not equivalent to the examples listed above
- Disk only, without prints

---

[1] The list below includes products known at this time to meet the minimum documentation specifications established for the compilation of National Register nomination documents. The list is not intended to be restrictive or comprehensive, and does not constitute, and shall not be taken as, endorsement by the State Historic Preservation Office of any of the specific products or manufacturers identified.

Page 5 of 6

The Disk:

BEST: CD-R - with patented Phthalocyanine dye and 24 Karat gold reflective layer.
- Examples:
  -Delkin's Archival Gold™ (also referred to as eFilm® Archival Gold)
  -MAM-A Gold ™(also know as Gold-On-Gold™)
  -Verbatim UltraLife™ Gold Archival Grade CD and DVD-R
Acceptable: CD-R or DVD-R
Not acceptable: CD-RW or DVD- RW

Labeling the Disk

BEST: Labels printed directly on the disk by way of inkjet or laser printers
Acceptable: Labeled using CD/DVD safe markers,
- Examples:
  -Sharpies™
  -Prismacolor®
Not acceptable: Ammonia or solvent based markers

VI. ADDITIONAL ITEMS

In addition to the items described in these guidelines, the SHPO may request additional documentation, depending on the nature and significance of a particular resource.

If you have any questions, please contact the Cultural Resources Management Specialist at 517-335-2721.

State Historic Preservation Office
Michigan Historical Center
702 West Kalamazoo Street
PO Box 30740
Lansing, MI 48909-8240

8/11

Page 6 of 6

United States Department of the Interior

OFFICE OF THE SECRETARY
Office of Environmental Policy and Compliance
Custom House, Room 244
200 Chestnut Street
Philadelphia, Pennsylvania 19106-2904

IN REPLY REFER TO:

January 9, 2011

9043.1
ER 11/1002

Mr. Bruce Olson
Project Manager
Environmental Projects Branch 2
Division of New Reactor Licensing
Office of New Reactors
U.S. Nuclear Regulatory Commission,
Washington, DC 20555-0001

Dear Mr. Olson:

The U.S. Department of Interior (Department) has reviewed the Draft Environmental Impact
Statement (DEIS) for the Combined License (COL) for the Enrico Fermi Unit 3 proposed by
Detroit Edison Company (DTE) (NUREG-2105). Fermi 3 is co-located with Units 1 and 2,
Monroe County, Michigan, on the shore of Lake Erie. These comments have been prepared
under the authority of the Fish and Wildlife Coordination Act (48 Stat. 401, as amended; 16
U.S.C. 661 et seq.), the Endangered Species Act of 1973 (ESA), as amended, and are consistent
with the intent of the National Environmental Policy Act of 1969 and the U. S. Fish and Wildlife
Service's Mitigation Policy.

Federally Threatened and Endangered Species

To facilitate compliance with Section 7(c) of the Endangered Species Act of 1973, as amended,
Federal agencies are required to obtain information from the U. S. Fish and Wildlife Service
(FWS) concerning any species, listed or proposed to be listed, that may be present in the area of
proposed action.

The DEIS identifies six federally-listed species in Monroe County, Michigan that may inhabit
the project area. The FWS is reserving substantive comments regarding federally listed species
until they are provided an opportunity to review the forthcoming biological assessment. At that
time, consultation pursuant section 7 of the ESA will continue. The construction of the
transmission lines will require a separate section 7 consultation as it is considered a separate
project by the Nuclear Regulatory Commission (NRC). The FWS recommends that the NRC not
issue a license for Fermi 3 until section 7 consultation has been completed.

Bald Eagles

There is a known bald eagle territory that overlaps DTE's FERMI 3 project boundary. As outlined in the FWS Bald Eagle Management Guidelines (http://www.fws.gov/midwest/eagle/guidelines/guidelines.html), the FWS recommends no construction activity within a buffer distance of 660 feet from any existing or recently existing nest if the proposed activity is visible from the nest and/or a resulting structure will be over three stories tall. Because the locations of proposed project-related construction activities appear to fall outside the recommended 660 foot nest buffer around the current active nest, the FWS has determined that this project, at this time, is unlikely to result in take of breeding eagles. This determination should only be considered valid as long as activities associated with the chosen project alternative continue to fall outside of the aforementioned 660 foot buffer around the current active eagle nest and there are no new eagle nests identified in the area.

It is worth noting that the breeding pair of eagles that occupy the nearby territory have constructed five nests in the last ten years (resulting in one new nest approximately every other year) on FERMI property, and have used all but one of them for nesting during that same time period. An unused nest was constructed in 2011 and is likely to be used for breeding at some point in the future. Because these eagles frequently relocate nest sites, and because the project start date may be one or several years down the road, it is very difficult to predict impacts to these eagles from this project. As such, FWS recommends that DTE remain in close contact with FWS Field Office in Michigan regarding changes in eagle nest locations. If a new nest were to be built, or an inactive nest be occupied in the future and project activities cannot be modified to avoid a potential disturbance, an eagle take permit may be necessary.

Additionally, since the project is located in the proximity of eagle foraging and roosting habitat both during breeding and in the winter, along with the above finding, the FWS encourages you to implement the following recommendations to further avoid impacting bald eagles:

- Minimize potentially disruptive activities (as outlined in the Guidelines) and development in the eagles' direct flight path between any known nests, roost sites and/or important foraging areas.
- Avoid loud, intermittent noises within one-half mile of known eagle nest locations during the breeding season and known eagle use areas when eagles are present .
- Protect and preserve potential roost and nest sites by retaining, when possible, mature trees and old growth stands within one-half mile of water.
- Employ industry-accepted best management practices to prevent birds from colliding with any lines, poles, and tower supports.
- Use pesticides, herbicides, fertilizers, and other chemicals only in accordance with federal and state laws.

Migratory Birds

The DEIS identifies several species of woodland and grassland bird species or their habitats that fall under protection of the Migratory Bird Treaty Act. Because the proposed project site very likely provides nesting habitat for migratory birds, we have concerns that the proposed project may also impact migratory birds. Under the Migratory Bird Treaty Act of 1918, as amended, it is unlawful to take, capture, kill, or possess migratory birds, their nests, eggs, or young. We recommend that removal of potential nesting habitat associated with the proposed project be completed before spring nesting begins or initiated after the breeding season has ended to avoid take of migratory birds, eggs, young, and/or active nests. Specifically, we recommend that no habitat disturbance, destruction, or removal occur between April 15 and August 15 to minimize potential impacts to migratory birds during their nesting season, but please be aware that some species may initiate nesting before April 15.

Wildlife Habitat

Approximately 197 acres of terrestrial wildlife habitat on the proposed Fermi 3 site will be disturbed and of that, 51 acres will be permanently lost. We would recommend DTE develop a wildlife management plan to compensate for the loss of wildlife habitat to be reviewed and approved by the FWS Field Office in Michigan. There will be approximately 130 acres of grassland-type habitat either permanently or temporarily lost due to the construction of Fermi 3 and associated appurtenances. The plan should include development of quality grassland habitat to offset the loss and to provide nesting habitat for grassland avian species (i.e., bobolink, Eastern meadowlark, savannah sparrow).

Wetlands and Aquatic Habitats

Approximately 34.5 acres of wetlands will be affected from the construction of Fermi 3. Of that, 27.7 acres will be temporarily disturbed and will be restored. Approximately 8.3 acres would be permanently lost at the site. To offset any wetland loss, DTE has developed an aquatic resource mitigation plan that includes restoring or enhancing approximately 82 acres of wetland offsite in the coastal zone of Western Lake Erie. The FWS agrees conceptually with the mitigation plan although according to the FWS's mitigation plan, coastal wetlands may be considered Category 1, with a goal of "no loss of existing habitat value." Therefore, the 0.80 acres of emergent coastal wetlands proposed to be impacted by the project should not lose any existing habitat value.

Pgs. 2-74, and 9-202: The information presented in the document on the Lake Erie fishery could be more thorough. USGS suggests that the Final EIS include the information available from the website: http://www.glsc.usgs.gov/_files/reports/2009LakeErieMonitoring.pdf

Pg. 2-121: The document does not indicate that the tubenose goby (*Proterorhinus semilunaris*) has been collected in Swan Creek. USGS suggests the Final EIS include the information on the tubenose goby available from the website:
http://nas.er.usgs.gov/queries/factsheet.aspx?SpeciesID=714

Pg. 9-153: The information presented in the document on the Lake Huron fishery could be more thorough. USGS suggests the Final EIS include the information available from these websites:
http://www.glsc.usgs.gov/_files/reports/2009LakeHuronDemersal.pdf
http://www.glsc.usgs.gov/_files/reports/2009LakeHuronPreyfish.pdf

Pg. 9-202, paragraph 3: The tubenose goby (*Proterorhinus semilunaris*) is not included in the list of nuisance species. USGS suggests the Final EIS include the tubenose goby as a nuisance species. A suggested reference can be found at:
http://nas3.er.usgs.gov/queries/CollectionInfo.asp?SpeciesID=714&HUCNumber=41000

3

Water Intake

DTE has proposed a closed circuit cooling system with a cooling basin cooling tower for Fermi 3. This closed system can significantly reduce the water use by 96 to 98%, and significantly reduce the impingement or entrainment of aquatic organisms. DTE has also proposed a through screen velocity of 0.5 ft/s or less under all operating conditions which should also reduce entrainment and impingement. The system also allows impinged organisms to be washed from the traveling screens to be directed back to Lake Erie via a fish return system. We laud these measures to reduce entrainment/impingement but the DEIS has not addressed impingement of diving ducks. There are water intake structures at other nuclear power plants in the Great Lakes where this has become a problem. Ducks may be attracted to the intake structures to feed on the guagga/zebra mussels that colonized the intake and the surrounding substrate. The DEIS has not stated the depth of the intake. The depth could be greater than a diving duck's diving capabilities but DTE should address this issue in the forthcoming FEIS.

Summary

The FWS will provide more substantive comments regarding federally listed threatened and endangered species after they are provided the opportunity to review the biological assessment (BA). In the DEIS, on page 5-21, it is stated that "the Review Team will prepare a BA prior to issuance of final EIS", at which time the U. S. Fish and Wildlife Service, East Lansing Field Office will review the BA. Wetland loss should be mitigated and any affected coastal wetland should not lose any exiting habitat value. A wildlife management plan should be developed and provided to the local FWS Office for review and comment. The impingement of diving ducks should be addressed in any forthcoming NEPA documents.

We appreciate the opportunity to provide these comments.

Sincerely,

Lisa Chetnik Treichel

Lisa Chetnik Treichel
Program Manger,
Land, Energy and Transit Projects

cc: Dave Larsen & Jeff Gosse, USFWS, Bloomington. MN

4

Appendix F

March 30, 2012

Mr. Scott Hicks, Field Office Supervisor
U.S. Fish and Wildlife Service
East Lansing Michigan Field Office
2651 Coolidge Road, Suite 101
East Lansing, MI 48823-6316

SUBJECT: SUBMITTAL OF THE BIOLOGICAL ASSESSMENT FOR THE PROPOSED
ENRICO FERMI NUCLEAR POWER PLANT, UNIT 3

Dear Mr. Hicks:

The U.S. Nuclear Regulatory Commission (NRC) staff is reviewing an application submitted by
Detroit Edison Company (Detroit Edison) for a combined license (COL) for construction and
operation of a new nuclear power plant, Enrico Fermi Nuclear Power Plant, Unit 3 (Fermi 3), at
its Enrico Fermi Atomic Power Plant (Fermi) site in Monroe County, Michigan, approximately 24
miles northeast of Toledo, Ohio, and 30 miles southwest of Detroit, Michigan. As part of the
review of this COL application, the NRC staff is preparing an environmental impact statement
(EIS) as required by Title 10 of the *Code of Federal Regulations* (10 CFR) Part 51, the NRC's
regulations that implement the National Environmental Policy Act of 1969 (NEPA), as amended.
The U.S. Army Corps of Engineers (USACE) is participating with the NRC in the preparation of
this EIS as a cooperating agency. The EIS includes an analysis of pertinent environmental
matters including those involving endangered or threatened species and impacts to fish and
wildlife. The NRC is submitting this letter and Biological Assessment as part of consultation
initiated on December 23, 2008 (ML083151398), under the Endangered Species Act of 1973
(ESA), as amended.

Detroit Edison submitted the application for a COL for Fermi 3 on September 18, 2008, pursuant
to NRC requirements in 10 CFR Part 52. The application is available through NRC's web-based
Agencywide Documents Access and Management System (ADAMS), which can be found at
http://www.nrc.gov/reading-rm/adams.html. The latest version of Detroit Edison's
Environmental Report (ER), which is Part 3 of the COL application, is listed under the accession
number ML110600498. The Fermi 3 COL application is also available on the Internet at
http://www.nrc.gov/reactors/new-reactors/col/fermi.html.

The Fermi site is located on approximately 1,260 acres along the western shore of Lake Erie.
The Fermi site currently has one operating boiling water reactor, Fermi Nuclear Power Reactor
Unit 2 (Fermi 2), which has the capacity to generate 1,089 megawatts of electricity. In addition,
there is one non-operating reactor, Fermi Nuclear Power Reactor Unit 1, which has been
defueled and is in the process of being decommissioned. Detroit Edison proposes to construct
one new nuclear unit adjacent to the existing facilities in areas that have been previously
disturbed and certain USACE regulated activities and structures associated with the project
would occur in waters of the United States and adjacent wetlands..

S. Hicks   - 2 -

The Fermi property is zoned as a Public Service District, which allows for power plant use. To support the Fermi 3 power plant, ITC *Transmission*, responsible for the transmission grid in southeastern Michigan, would have to build three new 345-kV transmission lines in a single corridor from the power plant to a substation in Milan, Michigan. The new transmission lines would be sited in portions of Monroe, southwest Wayne, and southeast Washtenaw Counties, Michigan.

The proposed cooling system is comprised of an intake structure and pipeline that extends into Lake Erie and circulating water systems throughout the plant. It allows for the loss of some water through heat dissipation into the atmosphere, and the discharge of water into Lake Erie. Approximately 34,000 gallons per minute would be withdrawn from Lake Erie during normal operations to make up losses from evaporation, drift, and blowdown. Waste heat would be dissipated to the atmosphere through a hyperbolic natural draft cooling tower. Blowdown from the cooling tower would be transported to an outfall that discharges into Lake Erie.

The NRC staff and USACE staff in its review, has evaluated the environmental impacts of construction and operation of Fermi 3, associated transmission lines, and alternatives, including alternative sites. NRC issued the draft EIS on October 28, 2011, and it is available on the NRC public Web site at http://www.nrc.gov/reactors/new-reactors/col/fermi.html.

To support the preparation of the NRC EIS on the proposed action, and to ensure compliance with Section 7 of the ESA, the NRC and USACE communicated with Burr Fisher of your office, by teleconference on July 26, 2011, and in onsite meetings on August 8-9, 2011. During the teleconference and meetings, NRC and USACE personnel discussed information on Federally listed species and critical habitat that may be in the vicinity of the proposed Fermi 3 site and the associated transmission line rights-of-way. The NRC and the USACE have prepared this Biological Assessment to support a joint consultation with the U.S. Fish and Wildlife Service (FWS) in accordance with the ESA, as amended.

Enclosed with this letter is the Biological Assessment that evaluates potential impacts to Federally listed species and habitats under the ESA. It also contains additional background information regarding the proposed Fermi 3 project.

S. Hicks   - 3 -

If you have any questions concerning the enclosed Fermi 3 Biological Assessment, please contact Mr. Bruce Olson, Environmental Project Manager at 301-415-3731 or by email at Bruce.Olson@nrc.gov. In his absence, please contact Mr. John Fringer at 301-415-6208, or by email at John.Fringer@nrc.gov.

Sincerely,

/RA/

Anthony H. Hsia, Chief
Environmental Projects Branch 2
Division of New Reactor Licensing
Office of New Reactors

Docket No.: 52-033

Enclosure:
As stated

cc w/o enclosure: See next page

United States Department of the Interior

FISH AND WILDLIFE SERVICE
East Lansing Field Office (ES)
2651 Coolidge Road, Suite 101
East Lansing, Michigan 48823-6316

IN REPLY REFER TO:

June 8, 2012

Anthony H. Hsia, Chief
United States Nuclear Regulatory Commission
Environmental Projects Branch 2
Division of New Reactor Licensing
Office of New Reactors
Washington, DC 20555-0001

RE: Endangered Species Act Section 7 Consultation for the Fermi 3 Nuclear Power Plant,
       Monroe County, Michigan

Dear Mr. Hsia:

We are in receipt of your cover letter dated March 30, 2012, with the accompanying biological
assessment (BA) for the construction and operation of a proposed nuclear power plant. Detroit
Edison (DTE) has submitted the application for a combined license (COL) for construction and
operation of the proposed Fermi Nuclear Power Plant, Unit 3 (Fermi 3) to be located on
approximately 1,260 acres along Lake Erie at the existing Enrico Fermi Nuclear Power in
Monroe County, Michigan.

The Fermi site currently has one operating boiling water reactor, Unit 2, and Unit 1 has been
defueled and is in the process of being decommissioned. The proposed construction of Fermi 3
is adjacent to the existing facilities in an area that has been previously disturbed. DTE has
identified the need for transmission line upgrades and three new transmission line corridors and a
separate switchyard. The siting area for the new transmission lines would include Monroe,
southwest Wayne, and southeast Washtenaw Counties, Michigan.

Your analysis addresses the potential effects of the project on the federally listed Indiana bat
(*Myotis sodalis*), Eastern prairie fringed orchid (*Platanthera leucophaea*), Karner blue butterfly
(*Lycaeides melissa samuelis*), Mitchell's satyr (*Neonympha mitchellii mitchellii*), American
burying beetle (*Nicrophorus americanus*), Northern riffeshell (*Epioblasma torulosa rangiana*),
rayed bean (*Villosa fabalis*), and snuffbox (*Epioblasma triquetra*) mussels. You have also
evaluated the potential effects of the project on the candidate Eastern massasauga rattlesnake
(*Sistrurus catenatus*).

You have determined the Fermi 3 project may affect but is not likely to adversely affect the Indiana bat, eastern prairie fringed orchid and the eastern massasauga rattlesnake. We concur with your determination that the construction and operation of the facility may affect, but is not likely to adversely affect the Indiana bat and eastern prairie fringed orchid.

Indiana Bat

In Michigan, summering Indiana bats roost in trees in riparian, bottomland, and upland forests from approximately April through October. Indiana bats may summer in a wide range of habitats, from highly altered landscapes to intact forests. Roost trees vary considerably in size, but those used by Indiana bat maternity colonies are typically greater than 9 inches dbh. Male Indiana bats have been observed roosting in trees as small as 3 inches dbh.

We concur that the proposed on-site actions are not likely to adversely affect the Indiana bat for the following reasons:

- There are currently no known locations of Indiana bats in Monroe County, and there is limited habitat on site.
- Given the small amount of potential habitat on-site, any effect on Indiana bats will be insignificant.

Eastern prairie fringed orchid

The eastern prairie fringed orchid (EPFO) may be found in lakeplain wet or wet-mesic prairie and will also persist in degraded prairie remnants, ditches, railroad rights-of-way, fallow agricultural fields, and similar habitats where artificial disturbance creates a moist mineral surface conducive to germination.

EPFO is not known to occur near the proposed project area. We concur that the proposed action is *not likely to adversely affect* the EPFO for the following reason:

- EPFO has not be observed on-site during the course of site surveys and suitable habitat is lacking.

Based upon this information, any effects on EPFO from this project would be discountable.

Eastern Massasauga Rattlesnake

The eastern massasauga rattlesnake occurs in a variety of wetland systems with adjacent upland habitat. Populations in southern Michigan typically use shallow, sedge- or grass-dominated wetlands, while those in northern Michigan prefer lowland coniferous forests, such as cedar swamps. This species requires open, sunny areas with scattered shade to assist with thermoregulation, but avoids heavily wooded or closed-canopy areas.

The species is currently a candidate under the Act and, as such, does not require consultation under section 7 of the Act. Although the Act does not extend protection to candidate species, we encourage and appreciate their consideration in project planning. Avoidance of unnecessary impacts to candidate species will reduce the likelihood that they will require the protection of the Act in the future.

Your BA also evaluated the effects to Karner blue and Mitchell's satyr butterflies, American burying beetle, the northern riffleshell, rayed bean and snuffbox mussels. You determined that the construction and operation of the facility will have "no effect" on these six federally-listed species. Although our concurrence with your "no effect" determination is not required under the Act, we are in agreement with your findings.

You have also made a determination of effects for the 29.4 miles of proposed transmission lines associated with the project. We are not able to concur with your effects determinations for the proposed transmission lines at this time. Your evaluation indicates that terrestrial and/or aquatic surveys for listed species will be conducted once the location of the transmission line corridors has been finalized. We will defer concurrence with your determinations until corridor locations are finalized and we have reviewed the results of future surveys. We also recommend that future surveys include those for the Indiana bat and for listed mussel species at stream crossings when the stream bottom is to be disturbed. Future consultation should be completed prior to submission of Michigan Department of Environmental Quality and/or the Army Corps of Engineers permit applications for stream crossings or wetland fill associated with the transmission line towers.

We appreciate this opportunity to provide comments and look forward to continued coordination in the future if necessary. Any questions should be directed to Mr. Burr Fisher at 517/351-8286 or burr_fisher@fws.gov.

Sincerely,

Acting for Scott Hicks
Field Supervisor

cc: MDNR, Wildlife Division, Lansing, MI (Attn: Lori Sargent)
       Detroit River International Wildlife Refuge, Grosse Ile, MI (Attn: John Hartig)

Biological Assessment

U.S. Fish and Wildlife Service

Enrico Fermi Unit 3
Combined License Application

Biological Assessment

U.S. Fish and Wildlife Service

Enrico Fermi Unit 3
Combined License Application

U.S. Nuclear Regulatory Commission Combined License Application
Docket No. 52-033

U.S. Army Corps of Engineers Permit Application
Permit Application No. 10-58-0011-P

Monroe County, Michigan

March 2012

U.S. Nuclear Regulatory Commission
Rockville, Maryland

U.S. Army Corps of Engineers
Detroit District

Abbreviations/Acronyms

ac   acre(s)

BMP   best management practice

CFR   *Code of Federal Regulations*
COL   combined construction permit and operating license
CWA   Clean Water Act

DA   Department of the Army

EIS   environmental impact statement
EMF   electromagnetic field
EPA   U.S. Environmental Protection Agency
ER   environmental report
ESA   Endangered Species Act of 1973, as amended

FR   *Federal Register*
ft   foot/feet
FWS   U.S. Fish and Wildlife Service

GEH ESBWR   General Electric Hitachi Economic Simplified Boiling Water Reactor
GEIS   generic environmental impact statement

kV   kilovolt(s)

m   meter(s)
MDNR   Michigan Department of Natural Resources
MDOT   Michigan Department of Transportation
mi   mile(s)
MNFI   Michigan Natural Features Inventory

NEPA   National Environmental Policy Act of 1969, as amended
NPDES   National Pollutant Discharge Elimination System
NRC   U.S. Nuclear Regulatory Commission

ROW   right-of-way

SESC	Soil Erosion and Sedimentation Control Plan
SWPPP	storm water pollution prevention plan
USACE	U.S. Army Corps of Engineers
USC	United States Code
USGCRP	U.S. Global Change Research Program

1.0 Introduction

The U.S. Nuclear Regulatory Commission (NRC) is reviewing an application from the Detroit Edison Company (Detroit Edison) for a combined construction permit and operating license (COL) to build one General Electric Hitachi Economic Simplified Boiling Water Reactor (GEH ESBWR) at the Enrico Fermi Atomic Power Plant (Fermi) site. The proposed NRC Federal action is the issuance, under the provisions of 10 CFR Part 52, of a COL authorizing the construction and operation of one new GEH ESBWR at the Fermi site. To support the Enrico Fermi Unit 3 (Fermi 3) power plant, ITC*Transmission*, responsible for the transmission grid in southeastern Michigan, would have to build three new 345-kV transmission lines in a single corridor from the power plant to a substation in Milan, Michigan. The Fermi 3 plant would be located adjacent to the existing Enrico Fermi Unit 2 (Fermi 2) plant within the 1,260-ac Detroit Edison Fermi site, located in Monroe County, Michigan. The Fermi site is approximately 30 mi southwest of Detroit, Michigan, approximately 24 mi northeast of Toledo, Ohio, and approximately 7 mi from the United States–Canada international border. Figure 1-1 depicts the 50-mi-radius region surrounding the Fermi site, and Figure 1-2 depicts the 7.5-mi-radius vicinity surrounding the Fermi site.

In addition to the COL application, Detroit Edison plans to apply for a U.S. Army Corps of Engineers (USACE) permit pursuant to Section 10 of the Rivers and Harbors Appropriation Act of 1899 and Section 404 of the Federal Water Pollution Control Act, also known as the Clean Water Act (CWA) for Fermi 3 work in navigable waterways and waters of the United States. The NRC is preparing an environmental impact statement (EIS) for the Fermi 3 project under the National Environmental Policy Act of 1969, as amended (NEPA). The USACE is cooperating with the NRC to ensure that the EIS is adequate to fulfill the requirements of USACE regulations; the CWA Section 404(b)(1) guidelines, which contain the substantive environmental criteria used by the USACE in evaluating discharges of dredged or fill material into waters of the United States; and the USACE public interest review process. The NRC and the USACE have prepared this biological assessment (BA) to support a joint consultation with the U.S. Fish and Wildlife Service (FWS) in accordance with the Endangered Species Act of 1973, as amended (ESA). The USACE permit decision will be made following issuance of the final EIS.

This BA examines the potential impacts of building and operating Fermi 3 at the Fermi site on Federally listed threatened or endangered species and species that are candidates for Federal listing pursuant to ESA Section 7(c). The BA also addresses Federally listed species and species that are proposed or candidates for Federal listing and could occur in the counties in Michigan that include the Fermi site or the proposed transmission system required to connect Fermi 3 to the electric grid (Table 1-1).

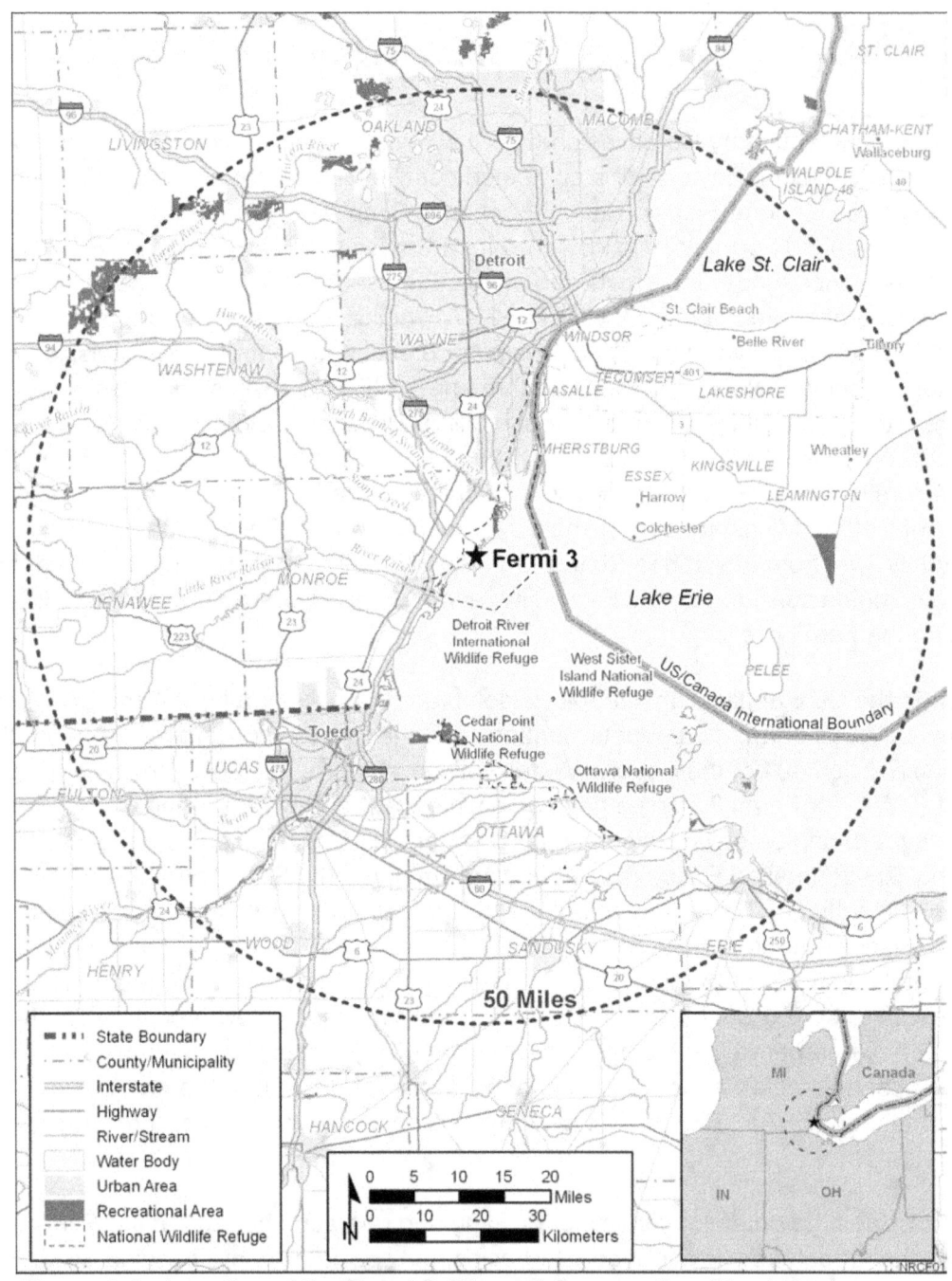

Figure 1-1. Location of the Fermi 3 Site and Surrounding 50-mi Region
(Source: Detroit Edison 2011a)

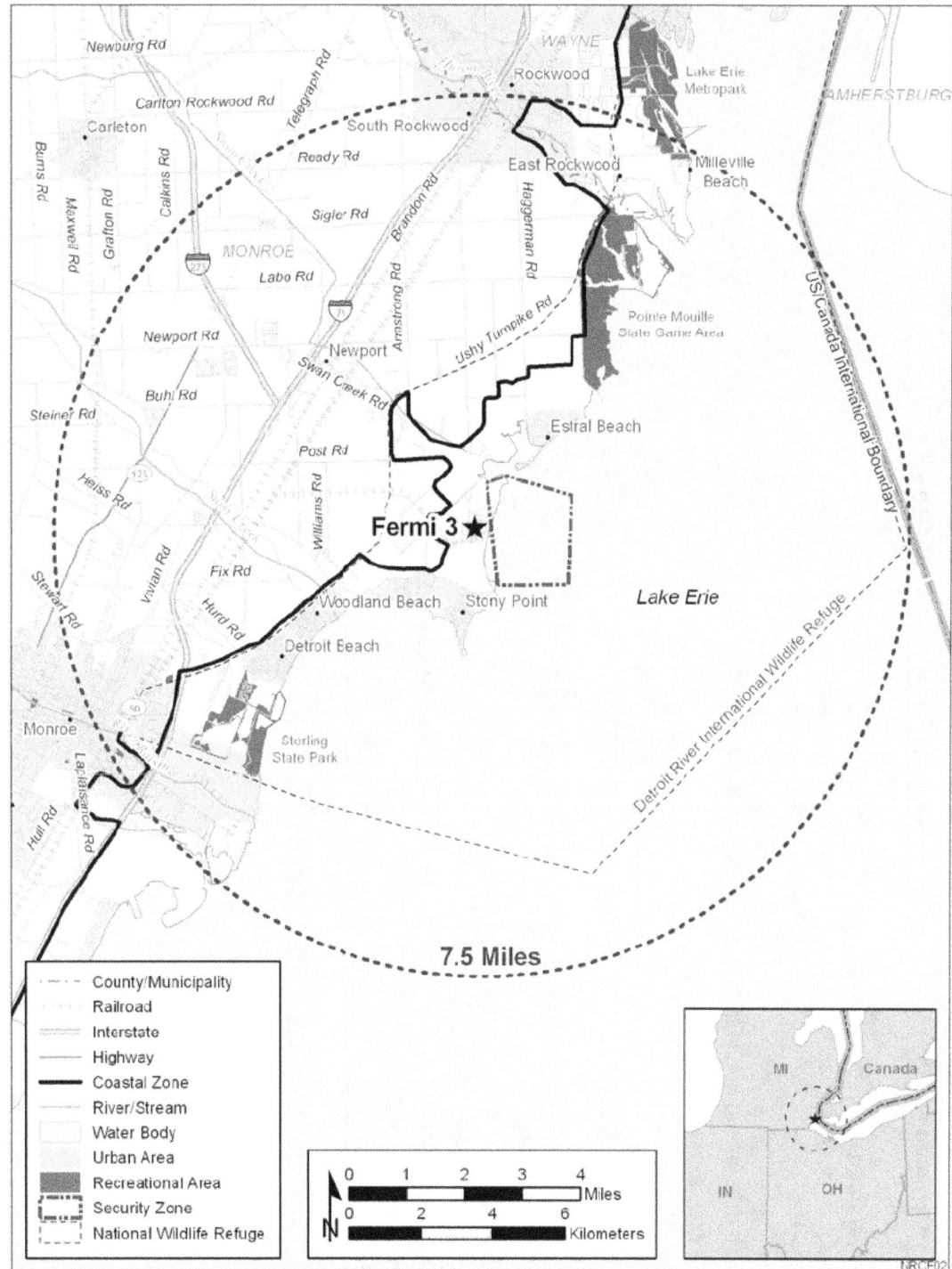

Figure 1-2. Fermi 3 Site and 7.5-mi Vicinity (Source: Detroit Edison 2011a)

Table 1-1. Federally Listed Threatened and Endangered Species and Species That Are Candidates for Federal Listing with a Potential to Occur in Counties In Which the Fermi Site and Proposed Transmission Lines Are Located

Common Name	Scientific Name	Legal Status[a]	County
Terrestrial Species			
Mammals			
Indiana bat	*Myotis sodalis*	E	Monroe, Washtenaw, Wayne
Reptiles			
Eastern massasauga rattlesnake	*Sitrurus catenatus catenatus*	C	Washtenaw, Wayne
Insects			
Karner blue butterfly	*Lycaeides melissa samuelis*	E	Monroe
Mitchell's satyr	*Neonympha mitchellii mitchellii*	E	Washtenaw
American burying beetle	*Nicrophorus americanus*	E	Washtenaw
Vascular Plants			
Eastern prairie fringed orchid	*Platanthera leucophaea*	T	Monroe, Washtenaw, Wayne
Aquatic Species			
Mollusks (Mussels)			
Northern riffleshell	*Epioblasma torulosa rangiana*	E	Monroe, Wayne
Rayed bean	*Villosa fabalis*	E	Monroe, Wayne
Snuffbox mussel	*Epioblasma triquetra*	E	Monroe, Washtenaw

Source: FWS (2009a).

(a) T = Federal threatened; E = Federal endangered; C = Federal candidate.

2.0 Fermi Site Description

The Fermi site is located in Monroe County, Michigan, along the shore of Lake Erie, approximately 30 mi southwest of Detroit, Michigan. The county's land use is mostly agricultural and rural, with some limited but growing residential areas. Areas of forests and wetlands are generally confined to property lines and along streams and shorelines (see Figure 2-1). The proposed Fermi 3 development area is located entirely within the current Fermi site boundary, just south and west of existing Fermi 2 facilities (see Figures 2-1 and 2-2). Part of the proposed Fermi 3 site was previously developed for the Fermi 1 and 2 Atomic Power Plants (Detroit Edison 2011a). Fermi 1 was last operated in 1972, is permanently shut down, and is being decommissioned. Fermi 2 is a licensed operating power plant. An aerial view of the Fermi site as it exists now is shown in Figure 2-3.

The project area (action area) consists of the Fermi site and the proposed transmission line corridor. Lake Erie borders the Fermi site on the east, Toll Road is located along the western boundary of the site, Swan Creek is located to the north, and Pointe Aux Peaux Road is located to the south. The entire Fermi site is relatively flat. Large areas of the site consist of developed land, but emergent wetlands, early successional habitats, forests, small quarry lakes, and ponds also are present. The locations of existing facilities at the Fermi site and facilities that would be developed for the proposed Fermi 3 project are shown in Figures 2-1 and 2-2.

The existing Fermi 2 unit will remain and continue to operate at the Fermi site and will not be affected by the proposed action. Fermi 2 uses two 400-ft-tall concrete natural draft cooling towers for heat dissipation (Figure 2-3). The cooling water intake from Lake Erie for Fermi 2 is located between two rock groins that extend into Lake Erie along the eastern edge of the site and is used to provide makeup water from Lake Erie for evaporation, drift, and blowdown losses. The Fermi 2 cooling water discharge is located along the shoreline of Lake Erie, north of Fermi 2 and east of the cooling towers (Figure 2-4).

2.1 Terrestrial Habitats – Vicinity and Site

The terrestrial communities found on the Fermi site and surrounding landscape are typical of the western shore of Lake Erie in the Lower Peninsula physiographic province and the Southern Lower Peninsula Ecoregion (MDNR 2005). The Fermi site is a mix of coastal emergent wetlands, developed areas, forests (including narrow coastal shoreline forests, lowland hardwoods, and woodlots), shrubland, and thickets. The surrounding landscape is generally flat and comprises a mix of agricultural fields, developed land, forested and emergent wetlands, and deciduous forests (Detroit Edison 2011a).

Figure 2-1. Fermi Site and Proposed Fermi 3 Facilities (Source: Detroit Edison 2011a)

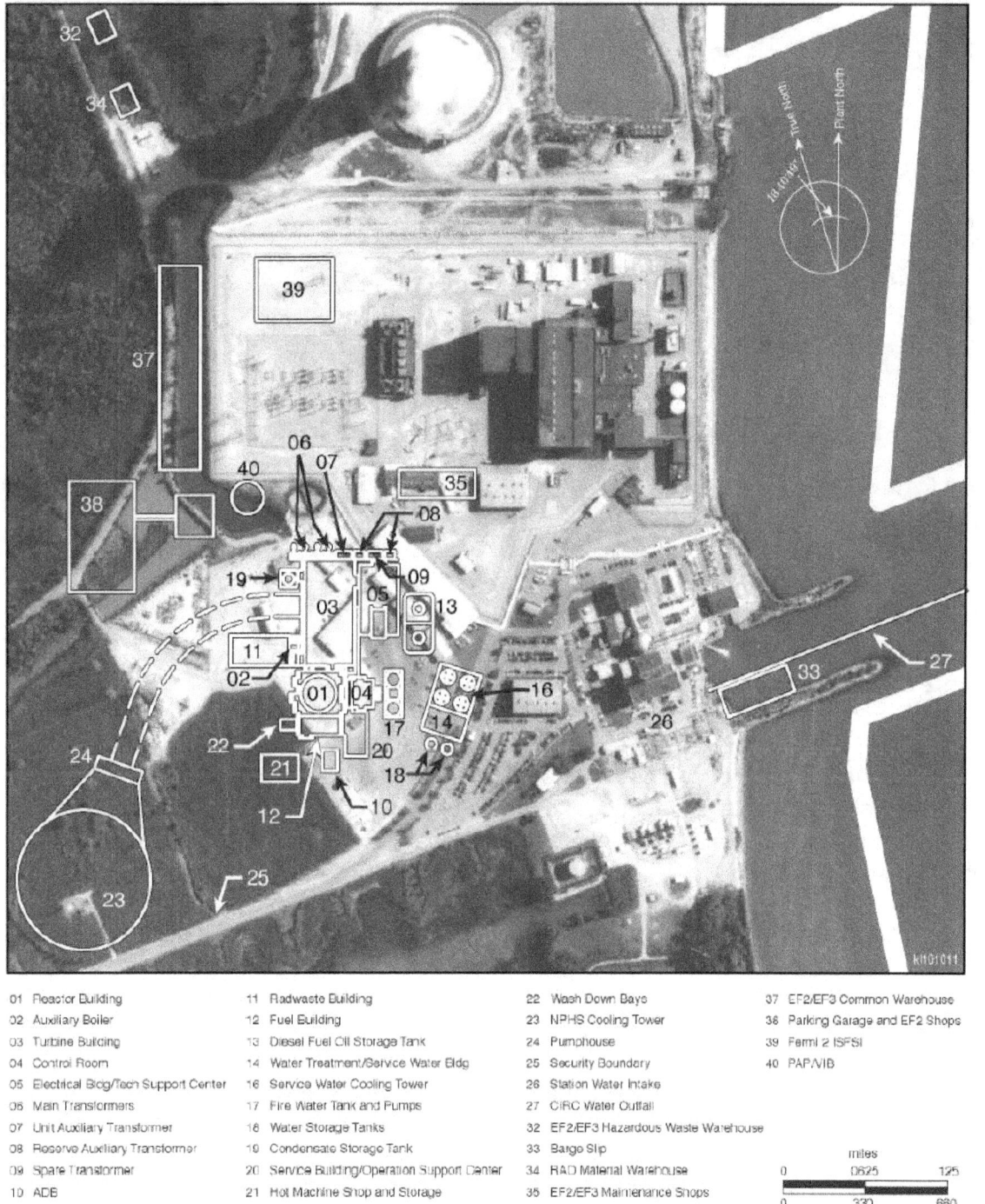

01 Reactor Building	11 Radwaste Building	22 Wash Down Bays	37 EF2/EF3 Common Warehouse
02 Auxiliary Boiler	12 Fuel Building	23 NPHS Cooling Tower	38 Parking Garage and EF2 Shops
03 Turbine Building	13 Diesel Fuel Oil Storage Tank	24 Pumphouse	39 Fermi 2 ISFSI
04 Control Room	14 Water Treatment/Service Water Bldg	25 Security Boundary	40 PAP/VIB
05 Electrical Bldg/Tech Support Center	16 Service Water Cooling Tower	26 Station Water Intake	
06 Main Transformers	17 Fire Water Tank and Pumps	27 CIRC Water Outfall	
07 Unit Auxiliary Transformer	18 Water Storage Tanks	32 EF2/EF3 Hazardous Waste Warehouse	
08 Reserve Auxiliary Transformer	19 Condensate Storage Tank	33 Barge Slip	
09 Spare Transformer	20 Service Building/Operation Support Center	34 RAD Material Warehouse	
10 ADB	21 Hot Machine Shop and Storage	35 EF2/EF3 Maintenance Shops	

Figure 2-2. Proposed Facilities at the Fermi Site (Source: Detroit Edison 2011b)

Figure 2-3. Aerial View of the Existing Fermi Site Looking North
(Source: Detroit Edison 2011a)

The most prevalent land cover types on the Fermi site are coastal emergent wetland, developed land, open water, woodlots, shrubland, and lowland hardwood. The surrounding area has similar cover types, except that coastal emergent wetlands and coastal forest are absent (Detroit Edison 2011a).

No surveys specifically designed to evaluate the Federally listed terrestrial species identified by the FWS, including species that are candidates for listing, have been conducted at the Fermi site. However, detailed terrestrial biological surveys of the Fermi site were conducted by Detroit Edison in July and October 2008 and May and June 2009 to support the EIS for the Fermi 3 project (Detroit Edison 2009a), and several previous wildlife and plant studies were conducted on the property. Detroit Edison conducted reconnaissance surveys of the Fermi site and vicinity between November 2006 and May 2008, and NUS Corporation examined the Fermi site between 1973 and 1974 prior to the development of Fermi 2 (NUS Corporation 1974; Detroit Edison 2011a). No Federally listed plants or animals or species that are candidates for Federal listing were observed during the surveys noted above (Detroit Edison 2009a, 2011a). No areas designated as critical habitat for Federally listed terrestrial species or species that are candidates for Federal listing exist at the Fermi 3 site.

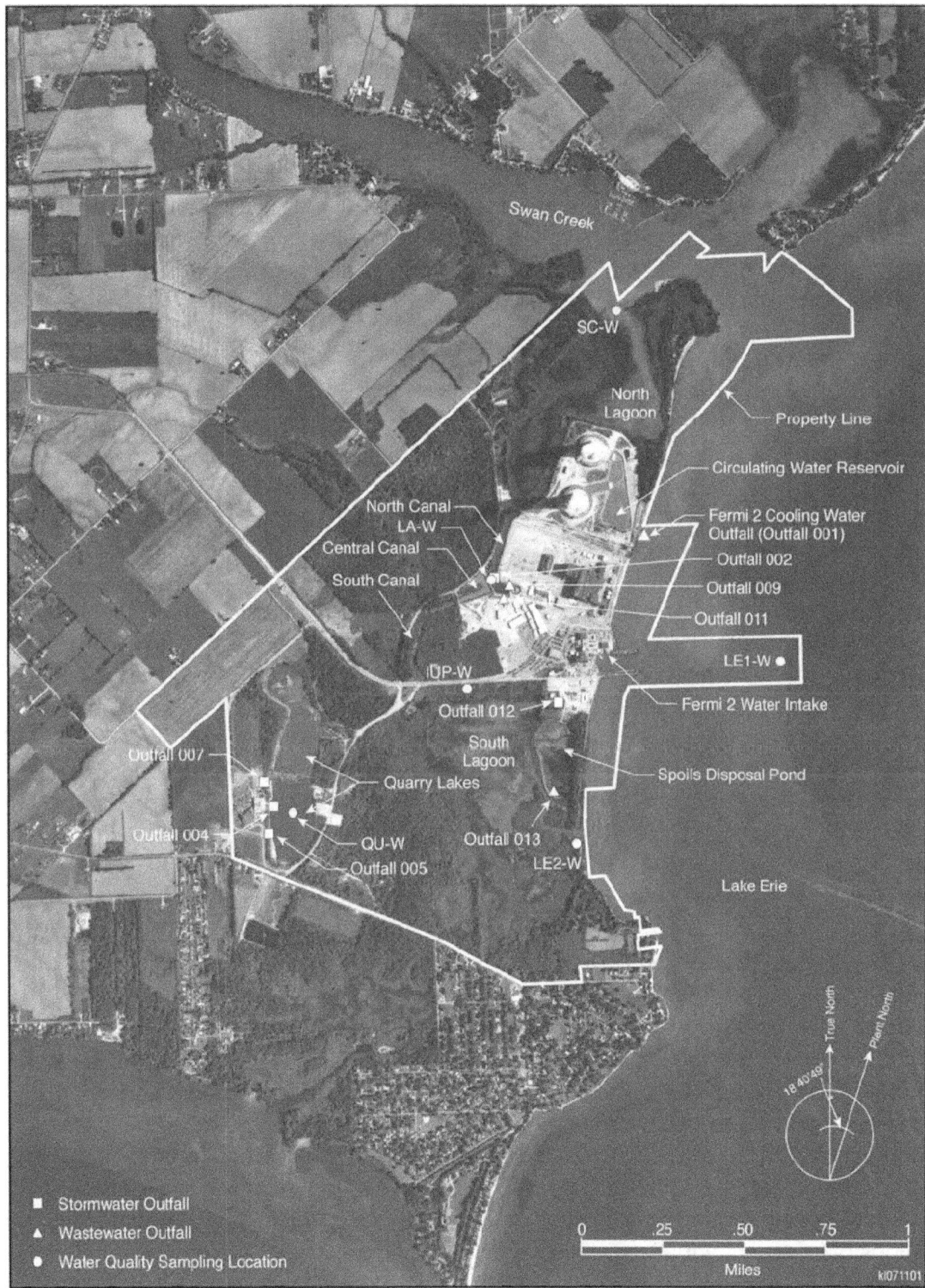

Figure 2-4. Surface Water Features, Discharge Outfalls, and Water Quality Sampling

Locations on the Fermi Site

A variety of wildlife species inhabit the forested, wetland, and open-water habitats on the Fermi site, including amphibians, reptiles, birds, and mammals. While the terrestrial wildlife species observed on the Fermi site are generally representative of the diverse but fragmented habitat types present on the site, the diversity of species is somewhat more limited than the habitat diversity might otherwise suggest. Although the habitat is diverse, habitat quality in the emergent marshes is compromised by the dense stands of common reed (*Phragmites australis*), which has low value as habitat for most species and aggressively competes with native plants that provide high-value habitat (Detroit Edison 2011a).

2.2 Aquatic Habitats – Vicinity and Site

The aquatic resources on the Fermi site and vicinity occur in a variety of natural and constructed freshwater features including (1) the circulating water reservoir, (2) overflow and discharge canals, (3) drainage ditches, (4) the onsite Quarry Lakes, (4) wetland ponds and marshes managed as part of the Detroit River International Wildlife Refuge (DRIWR), (5) Swan Creek, (6) Stony Creek, and (7) Lake Erie (Figure 2-4).

No surveys specifically designed to identify the Federally listed aquatic species identified by the FWS have been conducted at the Fermi site. However, detailed surveys of aquatic biota were conducted in a variety of aquatic habitats at the Fermi site by Detroit Edison from July 2008 through July 2009, to support the EIS for the Fermi 3 project (AECOM 2009), and several previous surveys have also been conducted in the vicinity of the Fermi site (e.g., MDEQ 1998; Gustavson and Ohren 2005; Francis and Boase 2007). No Federally listed aquatic species were observed during the surveys noted above. No areas designated as critical habitat for Federally listed aquatic species are present at the Fermi 3 site. Information about the aquatic habitats and biota associated with the various surface water features are provided in the following sections.

2.2.1 Circulating Water Reservoir

The circulating water reservoir, a component of the heat dissipation system associated with the operation of Fermi 2, provides cooling water for the circulating water system. The circulating water reservoir is located east of the Fermi 2 cooling towers in the northern portion of the developed part of the Fermi site (Figure 2-4). This manmade reservoir encompasses an area of approximately 5 ac, is approximately 20 ft deep, and is lined with clay. The circulating water reservoir is periodically treated with chemicals to inhibit excessive growth of vegetation and the production of aquatic organisms and does not provide habitat suitable for supporting significant populations of important aquatic species.

2.2.2 Overflow and Discharge Canals

One clay-lined canal, approximately 5 to 10 ft deep and 70 ft wide, originates in the central portion of the Fermi site (along the western edge of the developed portion of the site) and extends northward, where it connects with Swan Creek after passing through a marshy area known as the North Lagoon. This constructed canal is referred to as the north canal (Figure 2-4). The north canal was historically used as a cooling water discharge and overflow canal for operation of Fermi 1, but ceased being used when Fermi 1 was temporarily shut down in the mid-1960s. Currently, the Fermi site uses the canal as a permitted wastewater discharge (Outfall 009; Figure 2-4). Thirty fish species were captured in the overflow canal during surveys conducted in 2008; the most abundant species were bluegill (*Lepomis macrochirus*), pumpkinseed (*L. gibbosus*), emerald shiner (*Notropis atherinoides*), and gizzard shad (*Dorosoma cepedianum*) (AECOM 2009).

A second manmade canal, referred to as the south canal, originates in the central portion of the Fermi site and extends southward, where it flows into the South Lagoon (Figure 2-4). This canal is approximately 5 to 10 ft deep and 70 ft wide and serves as a drainage for wetland areas located west of the developed portion of the Fermi site. Twenty-eight fish species were collected in the discharge canal during surveys conducted in 2008; the most abundant species were goldfish (*Carrasius auratus*), common carp (*Cyprinus carpio*), bluegill, pumpkinseed, and golden shiner (*Notemigonus crysoleucas*) (AECOM 2009).

A third small water body is located between the overflow and discharge canals. This manmade feature, referred to as the central canal (Figure 2-4), is stagnant and has no connections to the overflow canal or the discharge canal. Thirteen fish species were collected in the central canal during surveys conducted in 2008; the most abundant species were bluegill, gizzard shad, largemouth bass (*Micropterus salmoides*), white crappie (*Pomoxis annularis*), green sunfish (*L. cyanellus*), and bluntnose minnow (*Pimephales notatus*) (AECOM 2009).

No Federally listed aquatic species were observed during sampling for fish or invertebrates in the overflow and discharge canals (AECOM 2009).

2.2.3 Quarry Lakes

The North and South Quarry Lakes (Figure 2-4) are located in the southwestern portion of the Fermi site. The two lakes are approximately 50 ft deep and, in total, cover an area of approximately 100 ac. The quarry lakes were created when water filled abandoned rock quarries that were used for site development and for development of Fermi 2 (Detroit Edison 1977) and have no surface water connection to other surface water habitats. The Quarry Lakes support a limited variety of aquatic species common to Lake Erie coastal marsh habitats. Nine fish species were collected in the Quarry Lakes during surveys conducted in

2008; the most abundant species were bluegill, gizzard shad, green sunfish, goldfish, and common carp (AECOM 2009). No Federally listed aquatic species were observed during sampling (AECOM 2009).

2.2.4 Wetland Ponds and Marshes

Portions of the Fermi site are managed as part of the DRIWR. These managed areas surround the developed portion of the Fermi site on the northern, western, and southern borders. The managed area encompasses approximately 656 ac, including coastal wetlands and palustrine wetlands, such as freshwater emergent wetlands and small lakes that are semi-permanently or seasonally inundated. A fisheries survey of coastal marsh managed areas was conducted in September 2005 by the Michigan Department of Natural Resources (MDNR) and FWS to document fish communities associated with Michigan waters of Lake Erie and to inventory fishery resources (Francis and Boase 2007). This survey used electrofishing and seining to sample four marsh complexes, one of which was the Swan Creek Estuary, near the northern extent of the Fermi site. A total of 38 species of fish from 13 families were collected at this sampling site. Species most common in the catch included gizzard shad, bluntnose minnow, mimic shiner (*Notropis volucellus*), bluegill, pumpkinseed, goldfish, and largemouth bass. Thirty-three fish species were collected during fishery surveys conducted near the mouth of Swan Creek in 2008. The most abundant species in those collections were gizzard shad, emerald shiner, bluegill, brook silverside (*Labidesthes sicculus*), pumpkinseed, and golden shiner (AECOM 2009). No Federally listed aquatic species were observed during sampling (AECOM 2009).

2.2.5 Swan Creek

Swan Creek is located on the northern boundary of the Fermi site (Figure 2-4). It originates approximately 12 mi to the northwest of the Fermi site as small streams and then flows south and east to where it enters Lake Erie. Land use adjacent to the Swan Creek drainage includes small residential communities and agricultural development. Swan Creek forms a freshwater estuary where it flows into Lake Erie. The aquatic habitat in this area is shallow, with large stands of submerged aquatic vegetation. Many areas along the shoreline support water lilies, cattails, common reed, and other emergent vegetation (Francis and Boase 2007; AECOM 2009). The benthic habitat associated with the area of Swan Creek adjacent to the Fermi site consists of sandy sediment interspersed with small pockets of gravel and flat stone (AECOM 2009).

Benthic macroinvertebrates were collected during eight sampling events from July 2008 through June 2009 near the location where water from the North Lagoon area enters Swan Creek (location SC-W in Figure 2-4; AECOM 2009). These collections were dominated by aquatic worms (Haplotaxida, 31 percent), small crustaceans (Amphipoda, 23 percent), and midge larvae

(Diptera, 19 percent), among others (AECOM 2009). Dreissenid mussels (zebra and quagga mussels) and pea clams were also present in the Swan Creek collections. A fisheries survey of the Swan Creek estuary was conducted in September 2005 by the MDNR and FWS, using electrofishing and seining to sample nine sites along Swan Creek ranging from approximately 0.5 to 2.5 mi from the Fermi site (Francis and Boase 2007). A total of 38 species from 13 families were collected at these sampling sites. Frequently encountered species included gizzard shad, bluntnose minnow, emerald shiner, mimic shiner, bluegill, pumpkinseed, goldfish, and largemouth bass (Francis and Boase 2007). Fish were also collected monthly from Swan Creek from July 2008 to June 2009 (excluding winter months) near the location where water from the North Lagoon area enters Swan Creek (location SC-W in Figure 2-4; AECOM 2009). Overall, the fish species encountered during these surveys were similar to those observed in the survey by Francis and Boase (2007). A total of 1790 fish, were represented in the samples, comprising 33 species, and dominant species included gizzard shad, emerald shiner, bluegill, brook silverside, and pumpkinseed (AECOM 2009). No Federally listed aquatic species have been reported from surveys conducted in Swan Creek.

2.2.6 Lake Erie

The Fermi site is situated along the shoreline of Lake Erie. Lake Erie would serve as the source of cooling water for Fermi 3 and would also receive cooling water discharge from the proposed unit. Consequently, aquatic habitats and organisms in Lake Erie in the vicinity of the Fermi site have the greatest potential for being affected by building and operation of Fermi 3. Lake Erie is divided into three basins based upon the bathymetry of the lake: the eastern basin, the central basin, and the western basin. Because the Fermi site is located on the shoreline of the western basin, this portion of Lake Erie is of the greatest concern relative to building and operation of the Fermi 3 unit.

Benthic invertebrates were sampled from two locations in Lake Erie just offshore from the Fermi site during 2008 and 2009 (AECOM 2009). One site was located in water approximately 3–5 ft deep and has a substrate that consists of mud and sand; this location is near the existing cooling water intake for Fermi 2 and the proposed location for the Fermi 3 intake. The benthic organisms collected at this site consisted primarily of various species of amphipods, dipterans, and tubificid worms (AECOM 2009). The second site, located in water approximately 1–4 ft deep at the southern end of the Fermi site near the South Lagoon, had a rocky substrate. Dominant taxa collected from this site included various species of ephemeropterans (mayflies), amphipods, dipterans, tubificid worms, mollusks (dreissenid mussels and Sphaerid clams), and water mites (AECOM 2009).

Fish were collected monthly from July 2008 to June 2009 (excluding winter months) at two sampling locations in Lake Erie just offshore from the Fermi site (AECOM 2009). One location was near the existing cooling water intake bay for Fermi 2, which is also the proposed intake location for Fermi 3. The other sampling location was approximately 0.5 mi south of the intake bay sampling location along the Lake Erie shoreline near the South Lagoon. The two locations differed in the types of aquatic habitat that were present and had comparatively different species richness and abundance. The intake location was located along a sand and gravel beach in the open waters of Lake Erie and had little or no structure that would provide cover or spawning features. The South Lagoon location was near sand and gravel shoreline areas, as well as vegetated shoreline areas that could provide cover and spawning areas for some fish species. In addition, the South Lagoon location was near the mouth of the drainage area for the South Lagoon, which has extensive aquatic vegetation; fish within that drainage can move freely from the lagoon out into the main body of the lake. Overall, 5765 individual fish, comprising 40 species, were collected from the two Lake Erie sampling locations (AECOM 2009). The most abundant species encountered in those collections were gizzard shad, goldfish, white perch (*Morone americana*), emerald shiner, spottail shiner, and bigmouth buffalo (*Ictiobus cyprinellus*) (AECOM 2009).

Additional data on fish species that occur in the waters of Lake Erie near the Fermi site are provided by entrainment and impingement sampling. The rates at which fish eggs and fish larvae were entrained by the existing cooling water intake of Fermi 2 were measured from July 2008 through July 2009, excluding months of December through February, when ice cover was present and it was anticipated that spawning by fish would be at minimum levels (AECOM 2009). Entrainment rates (fish eggs plus larvae per unit volume of water) ranged from $4.82/m^3$ in July 2009 to $0.00/m^3$ in November 2008 and March 2009. The average annual entrainment rate for all species collected from July 2008 through July 2009 was $0.98/m^3$. Of the 12 fish species identified in entrainment samples, the species with the highest annual entrainment rates included gizzard shad, emerald shiner, bluntnose minnow, and yellow perch (AECOM 2009). In general, fish species entrained during the 2008–2009 study (AECOM 2009) were similar to those captured during a previous entrainment study (Lawler, Matusky, and Skelly Engineers 1993) conducted at the Fermi site from October 1991 to September 1992. The most abundant larval fish taxa entrained during the earlier study included species in the family Cyprinidae, gizzard shad, species in the family Clupeidae, and white perch; the most abundant taxa for fish eggs in entrainment samples included Cyprinidae and Percidae.

Impingement data collected from 1991 to 1992 from the Fermi 2 intake indicated that the dominant species impinged was the gizzard shad, which accounted for 71.5 percent of the estimated total number of individual fish impinged during the study period. White perch was the second most abundant species impinged (6.8 percent of the estimated total). Third, fourth, and fifth species ranked by the estimated number of individuals affected were the rock bass, freshwater drum, and emerald shiner. An additional study to estimate impingement rates at the Fermi 2 intake was conducted from August 2008 through July 2009. During that period, gizzard shad accounted for approximately 39 percent, emerald shiner accounted for approximately 29 percent, and white perch accounted for approximately 10 percent of the total estimated numbers of fish impinged at the plant (AECOM 2009). Overall, it is estimated that 3102 individual fish were impinged by the Fermi 2 cooling water intake during the 2008–2009 sampling period. No Federally listed aquatic species were observed during the impingement study.

2.3 Terrestrial Habitats – Transmission Line Corridors

To deliver the power generated by Fermi 3, existing transmission line corridors would need to be upgraded and new corridors, complete with transmission lines and substations, would need to be developed. The proposed and existing transmission line corridor routes are indicated in Figure 2-5.

The need for additional transmission towers and additional right-of-way (ROW) width would be evaluated by ITC*Transmission* when designing the Fermi 3 connection in the future. Detroit Edison expects that Fermi 3 would require three 345-kilovolt (kV) lines in a single 300-ft-wide corridor extending north from the Fermi site and then west to the Milan Substation, for a total distance of about 29.4 mi. The anticipated route crosses portions of Monroe, Wayne, and Washtenaw counties (Figure 2-5).

The first segment (approximately 18.6 mi) follows existing 345-kV lines. North from the Fermi site, this segment of the proposed transmission line corridor follows existing Fermi 2 transmission lines to a point just east of I-75. From there, it runs west and north following other existing non-Fermi lines. Detroit Edison expects that the new transmission infrastructure for this first segment would fit within the existing corridor width already established for the existing transmission lines (Detroit Edison 2011a). In addition, reconfiguration of existing conductors may allow the use of existing transmission infrastructure in places.

The final 10.8 mi of the route would cross agricultural land, forest, and rural residential land. Although ITC*Transmission* has already established this segment of ROW, it would require clearing vegetation for a new ROW and erecting new towers and stringing transmission lines.

The route crosses vegetative cover types similar to those on the Fermi site and its vicinity, but there are no areas of coastal emergent wetlands or coastal forest along the transmission line corridor (Detroit Edison 2011a). ITC*Transmission* has not conducted systematic terrestrial and aquatic surveys for the Fermi 3 lines. Instead, the BA relies on information about the possible occurrence of endangered or threatened species in counties crossed by the transmission lines from FWS records (FWS 2011a) and the Michigan Natural Features Inventory (MNFI) (MNFI 2007a). The route does not cross any areas designated as critical habitat for endangered species.

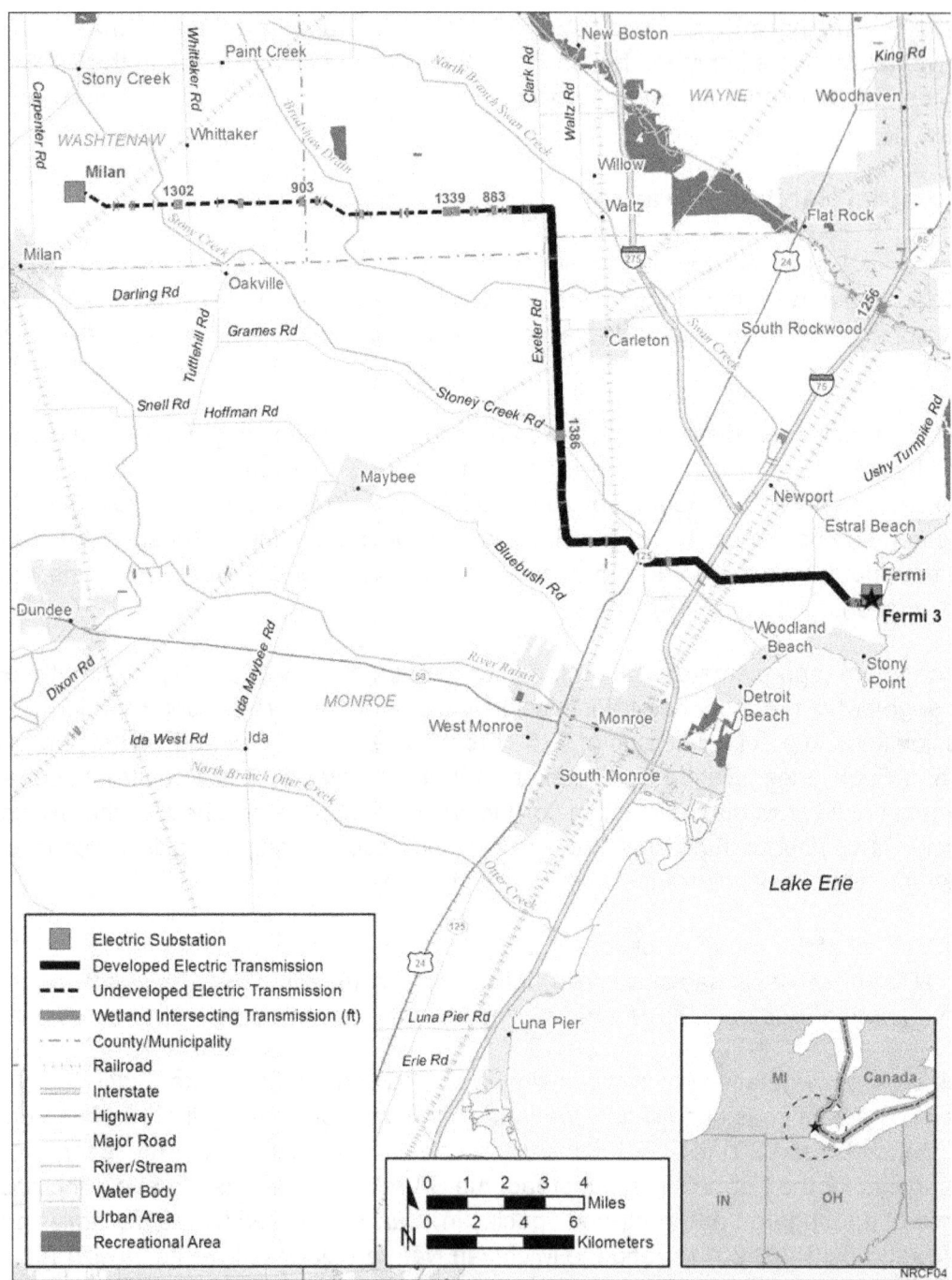

Figure 2-5. Proposed (Undeveloped) and Existing (Developed) Transmission
Line Corridors for Fermi 3

2.4 Aquatic Habitats – Transmission Line Corridors

Aquatic habitats within or adjacent to the corridor for the transmission lines needed to deliver power from Fermi 3, and identified in Detroit Edison's ER (Detroit Edison 2011a), include several small streams and numerous small drains that transport runoff water from agricultural areas. The undeveloped ROW where new transmission lines will be constructed crosses nine small streams and agricultural drains, but does not cross any lakes, ponds, or reservoirs (Figure 2-5). Stony Creek, which is crossed by the previously developed eastern portion of the assumed transmission line route and would be crossed again by the currently undeveloped portion of the assumed transmission line route, is the largest stream crossed by the transmission line corridor and is described below. Because of the small size of the remaining streams and agricultural drainages present along the presumed transmission line path, information regarding the aquatic species present in these water bodies is not available. There are no areas containing designated critical habitat along the transmission corridor (Detroit Edison 2011a). ITC*Transmission* has not conducted systematic aquatic surveys for the Fermi 3 lines. Instead, the BA relies on information about the possible occurrence of endangered or threatened species in counties crossed by the transmission lines from FWS records (FWS 2011a) and the MNFI (MNFI 2007a).

Although the Fermi site lies entirely outside of the Stony Creek watershed, some transmission line facilities associated with the proposed Fermi 3 development could cross streams located within the Stony Creek watershed. Stony Creek is located generally to the west of the Fermi site in Washtenaw and Monroe Counties, Michigan, and drains directly into the western basin of Lake Erie at a location approximately 3 mi southwest of the Fermi site boundary. Overall, Stony Creek is about 35 mi long and is supported by many more miles of smaller tributaries that comprise the Stony Creek Watershed.

Some biological data have been collected from Stony Creek and its tributaries. The Stony Creek Watershed Project has performed studies focusing on water quality, nutrients, and indicator species, although the majority of the data from these studies were not collected near the Fermi site. A macroinvertebrate survey to assess water quality was conducted in 2004 at several sampling sites along Stony Creek. The nearest sampling site was located approximately 2.5 mi south-southwest of the Fermi site. Data about various hydrological parameters were collected in addition to the macroinvertebrate samples (Gustavson and Ohren 2005). Fish surveys conducted in portions of Stony Creek located in Monroe County during 1997 indicated that the fish community in Stony Creek was dominated by taxa that are tolerant of degraded water quality conditions, although the fish community was rated as acceptable (MDEQ 1998). Dominant species found to be present included green sunfish, rock bass (*Ambloplites rupestris*), common carp, and blackside darter (*Percina maculata*)

(MDEQ 1998). No Federally listed aquatic species have been reported from surveys conducted in Stony Creek.

3.0 Proposed Federal Actions

The proposed NRC Federal action is the issuance, under the provisions of 10 CFR Part 52, of a COL authorizing the construction and operation of one new GEH ESBWR at the Fermi site. The proposed USACE Federal action is the issuance of a permit pursuant to the CWA and Rivers and Harbors Act of 1899 to authorize work that could affect waters of the United States, including jurisdictional wetlands.

Prerequisites to certain NRC-authorized construction activities include, but are not limited to, documentation of existing site conditions for the Fermi 3 site and acquisition of the necessary permits (e.g., COL, local building permits, a National Pollutant Discharge Elimination System [NPDES] permit [40 CFR Part 122], a CWA Section 404 permit, a General Stormwater Permit, and other state and local permits). After these prerequisites are completed, planned construction activities could proceed and would include all or some of the activities pursuant to 10 CFR Part 50.10(a)(1). Following construction, the planned operation of the new reactor would be authorized if the Commission finds, under 10 CFR 52.103(g), that all acceptance criteria in the COLs have been met.

In a final rule dated October 9, 2007 (NRC 2007), the NRC limited the definition of "construction" to the activities that fall within its regulatory authority in 10 CFR Part 51.4. Many of the site preparation activities associated with building a nuclear power plant are not part of the NRC action to license the plant. Activities that are associated with building the plant but are not within the purview of the NRC action are grouped under the term "preconstruction." Preconstruction activities include clearing and grading, excavating, erecting support buildings and transmission lines, and other associated activities. These preconstruction activities may take place before the application for a COL is submitted, during the review of a COL application, or after a COL has been granted. Although preconstruction activities are outside the NRC's regulatory authority, many of them are within the regulatory authority of local, State, or other Federal agencies. The distinction between construction and preconstruction is not carried forward in this biological assessment; both are jointly discussed and are generally referred to as "building."

The USACE regulatory program was originally established pursuant to the Rivers and Harbors Appropriation Acts of 1890 (superseded) and 1899 (33 USC 401 et seq.). Various sections establish permit requirements to prevent unauthorized obstruction or alteration of any navigable water of the United States. The most frequently exercised USACE authority is contained in Section 10 (33 USC 403). This section covers building, excavation, or deposition of materials in, over, or under such waters, or any work that would affect the course, location, condition, or capacity of those waters. In 1972 and 1977, amendments to the CWA added Section 404 authority, which authorizes the USACE to issue permits for the discharge of material into waters of the United States at specified disposal sites. Selection of such sites must be in accordance with guidelines developed by the U.S. Environmental Protection Agency (EPA) in conjunction with the Department of the Army (DA). These guidelines are known as the 404(b)(1) Guidelines for the specification of disposal sites for dredged or fill material. The discharge of all other pollutants into waters of the United States is regulated under Section 402 of the CWA.

Based on their habitat affinities and life-history characteristics, some protected terrestrial and freshwater species could be affected by building and operation activities associated with the Fermi 3 project:

- Terrestrial, including wetlands
    - Building activities
        - Onsite clearing, grading, and other site-preparation and building activities
        - Clearing for expansion of existing transmission line corridors or temporary workspaces
        - Clearing for new transmission line corridors
        - Installation of new or upgraded transmission lines and towers
    - Operation
        - Vegetation control in transmission line corridors
        - Transmission line repairs or upgrades

- Aquatic
    - Building activities
        - Terrestrial habitat disturbance, including wetlands and floodplains, on and in the vicinity of the site, and within/along existing and new transmission line corridors where such could impact water bodies (e.g., via erosion/sedimentation)
        - Aquatic habitat disturbance (e.g., dredging or placement of facilities in aquatic habitats)
    - Operation
        - Cooling water intake and discharge system
        - Transmission line ROW management
        - Introduction of contaminants (due to biocide and other water treatments, cooling tower blowdown, etc.)
        - Dredging to maintain the intake bay and barge slip areas

3.1 Impacts from Building and Operation on Site

The impacts from the proposed building and operation on onsite terrestrial and aquatic resources were assessed, as described in the following sections.

3.1.1 Terrestrial Species

Impacts on terrestrial resources, including wetlands, from building of Fermi 3 would include loss of habitat (temporary and permanent), increased human presence, increased traffic and noise, avian collisions with cranes and other tall construction equipment, the presence of outdoor lighting, and fugitive dust. Habitat losses would likely displace relatively mobile wildlife, while less-mobile wildlife could be destroyed. Mortality is expected to be limited mostly to the least-mobile wildlife, mainly small, slow-moving, burrowing, and cavity-dwelling species. However, increased mortality of more mobile species may result from increased traffic. Land clearing during nesting could temporarily depress local migratory bird productivity. Although nearby undisturbed forest and wetland habitat would be available to receive displaced animals, displaced wildlife would increase competition for limited resources, disrupt established territories, and cause increased predation and decreased fecundity. These conditions could lead to a temporary, localized reduction in population size for some species. After building activities have been completed, species that can adapt to disturbed or developed areas may recolonize affected areas where suitable habitat remains, is replanted, or allowed to regenerate.

The footprint of disturbance would encompass approximately 197 ac within the Fermi site (Detroit Edison 2011a). Approximately 9.4 ac of wetland on the Fermi site would be permanently lost, and approximately 23.7 ac would be temporarily disturbed during building. Approximately 9.4 ac of forest would be permanently lost, and approximately 11.1 ac would be temporarily disturbed during building. Approximately 8.4 ac of the permanently lost forest is recently regenerated cover on fill created during building of Fermi 2. Only about 1 ac of Lake Erie shoreline forest would be permanently disturbed. Section 4 describes Federally listed terrestrial species or species that are candidates for listing that may occur in or near the Fermi site or the proposed transmission line corridors.

Detroit Edison has stated its commitment to compliance with USACE Section 404 permit conditions and implementation of associated plans, including a Soil Erosion and Sedimentation Control (SESC) Plan, Storm Water Pollution Prevention Plan (SWPPP), and Compensatory Mitigation Plan, to provide adequate environmental protection (Detroit Edison 2011b). Best management practices (BMPs) would also be in place to address unavoidable disturbances. All building activities would be performed by Detroit Edison in compliance with applicable Federal, State, and local laws, regulations, and permit requirements (Detroit Edison 2011b).

Operating the proposed Fermi 3 plant is likely to have minimal potential impacts on vegetation, birds, and terrestrial, wetland, and shoreline habitats. Operations that could affect terrestrial resources are generally associated with the cooling system, transmission system, or traffic. Operation of the cooling towers transfers heat to the atmosphere in the form of water vapor and can result in icing, fogging, increased humidity, increased noise levels, and the deposition of dissolved solids from cooling-tower drift (NRC 1996). According to Detroit Edison, the maximum predicted annual salt deposition rate at any receiving location is 1 kg/ha/mo (Detroit Edison 2011a). This value is much lower than the NRC-acceptable levels of total dissolved solids and is not considered damaging to plants (NRC 2000). Therefore, impacts associated with operation of the cooling tower are expected to be negligible on vegetation, both on the Fermi site and in the vicinity.

Tall structures introduce a risk of avian collision mortality, but impacts on bird populations from avian collisions would be expected to be minimal. The Generic Environmental Impact Statement (GEIS) for license renewal (NRC 1996) concludes that effects of bird collisions with existing cooling towers are unlikely to threaten the stability of local populations or result in a noticeable impairment of the function of a species within local ecosystems.

Increased vehicular traffic could increase mortality of some wildlife species, particularly slower moving animals such as reptiles.

3.1.2 Aquatic Species

Activities related to building of Fermi 3 that could affect aquatic habitats include (1) building of a new intake structure, (2) building of a cooling water discharge structure, (3) rehabilitation of the existing barge slip, (4) building of a parking structure and a warehouse, and (5) dewatering of the Fermi 3 excavation area. Aquatic habitat features that could be directly affected by building Fermi 3 include Lake Erie and the north, central, and south canals on the Fermi site. Ground-disturbing activities that lead to soil erosion during site preparation and building of the new unit could result in adverse effects on water quality in water bodies on or adjacent to the Fermi site, including Lake Erie, the north and south canals, Swan Creek, and wetlands. Dewatering of the excavation area for Fermi 3 could result in lowering of groundwater levels that, in turn, could affect the level of surface water in the onsite North and South Quarry Lakes. In addition, during building of new transmission lines, there is a potential to affect habitat in streams that would be crossed in Monroe, Washtenaw, and Wayne Counties. This subsection evaluates impacts that could occur on aquatic resources on or in the vicinity of the Fermi site during building of Fermi 3 or during building of associated transmission lines. Building-related activities that could affect wetlands, including those within areas managed as part of the DRIWR, are described in Section 3.1.1 of this BA. As discussed in Section 2.2, drainage ditches and the circulating water reservoir on the Fermi site do not provide suitable aquatic habitat to support significant populations of aquatic organisms. Consequently, there would be little to no building-related impact on aquatic resources within these surface water features.

3.2 Impacts from Building and Operation in Proposed Transmission Line Corridors

The transmission lines serving Fermi 3 would be owned and operated by ITC*Transmission*. Detroit Edison would not control the development or operation of the new transmission lines. Accordingly, the following discussion is based on publicly available information and reasonable expectations of how ITC*Transmission* would proceed, based on standard industry practice.

3.2.1 Terrestrial Species

3.2.1.1 Building

Building Fermi 3 would necessitate development of three new transmission lines in an assumed 300-ft-wide corridor from the Fermi site to the Milan Substation, a distance of approximately 29.4 mi. The first 18.6 mi (going west and north from Fermi) would be installed alongside the 345-kV lines that are already in place (see Figure 2-5). This 18.6-mi portion of the transmission line would be created largely by the reconfiguration of conductors on existing towers within the transmission ROW, but placement of additional transmission infrastructure may be necessary (Detroit Edison 2011a). A majority of the 18.6-mi portion of the route would cross large crop fields and would result in minimal impacts on habitat and wildlife.

The 10.8-mi portion of ROW between the existing transmission ROW and the Milan Substation would run through forests, rural residential areas, and agricultural fields. For the purpose of this BA, the 10.8-mi portion of the proposed route is presumed to have a ROW that is 300 ft wide. To accommodate erection of new transmission towers, including installation of steel poles, footings, and conductors along this portion of the corridor, Detroit Edison has indicated that acquisition and clearing of additional land adjacent to the existing ROW may be necessary for laydown and other building purposes (Detroit Edison 2011a).

The Milan Substation would probably be expanded from its current size of 350 ft by 500 ft to an area approximately 1000 ft by 1000 ft to accommodate the three new transmission lines from Fermi 3 (Detroit Edison 2011a). This expansion would encroach onto maintained grass and agricultural areas.

The exact locations (routes) for the new ROWs have not yet been finalized by ITC*Transmission*. Thus, the routes and corridor boundaries shown in Figure 2-5 are considered provisional and subject to change (Detroit Edison 2011a). Field surveys for Federally listed threatened and endangered species or species that are candidates for Federal listing have not yet been conducted in the proposed corridors. No Federally listed terrestrial species or species that are candidates for Federal listing are known to occur in the affected or directly adjoining habitats, but several Federally listed terrestrial species could potentially use the corridor and adjoining habitats (MNFI 2007a; FWS 2011a). Wetland delineation surveys have not yet been conducted to determine the precise locations and extent of wetlands.

Development of the western 10.8 mi of transmission line corridor would affect approximately 415 ac. Approximately 244 ac of forest, including approximately 74 ac of forested wetlands, would be cleared of trees and other woody vegetation and planted with grass to accommodate the proposed three new 345-kV transmission lines (Detroit Edison 2011a). Other land uses within the proposed transmission line corridor include approximately 9 ac of emergent wetlands, 10 ac with grass or herbaceous cover, and approximately 135 ac of cropland, pasture, and hayfield. Most of the forested wetlands would be converted in the long term to scrub-shrub or emergent wetland types by controlling the regrowth of trees and other woody vegetation during maintenance of the corridor. The total potential permanent impact on wetlands from installation of the towers is expected to be approximately 0.5 ac (Detroit Edison 2011a).

Activities associated with building the new transmission lines would include clearing land, erecting new poles or towers, stringing new conductors, and upgrading existing transmission lines. Figure 2-5 shows the proposed routing for the three new lines in the transmission line corridors.

Impacts on wildlife and habitat from transmission line development would be reduced to the extent practicable by using existing transmission towers and ROWs for approximately 18.6 of the 29.4 mi of its length. Most large wildlife species present are expected to be sufficiently mobile and would temporarily move out of the way to avoid building activity, but smaller ground- and cavity-dwelling animals would be more vulnerable to mortality from land clearing. Wildlife species that favor disturbed vegetation communities would be expected to benefit and use the newly cleared ROW following erection of the transmission lines. The impact on terrestrial wildlife resources would therefore be minor.

ITC*Transmission* would have to mitigate unavoidable permanent wetland impacts to comply with Federal and State regulations. ITC*Transmission* would likely design mitigation measures in consultation with applicable regulatory agencies, including the USACE and MNDR, prior to submitting their permit applications (Detroit Edison 2011a). Prior to applying for permits from USACE and MDNR, ITC*Transmission* would likely have to delineate wetlands and conduct targeted surveys for Federally listed species. ITC*Transmission* could use that information to identify span and tower locations that minimize potential impacts on wetlands and other important habitats. ITC*Transmission* would at that time identify specific locations of towers, construction access routes, and material storage areas.

3.2.1.2 Operation

The potential effects on terrestrial ecological resources from transmission line operation would result mostly from vegetation maintenance. The GEIS for license renewal (NRC 1996) concludes that once a transmission line corridor has been established, the impacts on wildlife populations from continued ROW maintenance are not significant.

Effects on wildlife from the transmission lines are expected to be minor and to be limited to bird collisions with towers and conductors. Section 4.5.6.2 of the GEIS for license renewal (NRC 1996) concludes that bird collisions during operation of transmission lines do not cause long-term reductions in bird populations. The GEIS (NRC 1996) also concludes that the impacts of electromagnetic fields (EMFs) on terrestrial flora and fauna are not significant at operating nuclear power plants, including transmission line systems with variable numbers of power lines. On this basis, it is concluded that the incremental impacts of EMF due to possible additions of new power lines for Fermi 3 would be minimal.

Therefore, the review team concludes that the potential effects of transmission line maintenance in existing and new transmission line corridors would not likely adversely affect the Federally listed terrestrial species, including species that are candidates for Federal listing, identified in Table 1-1.

3.2.2 Aquatic Species

A short length (less than 1 mi) of new transmission line corridor would be developed on the Fermi site to transmit power from the Fermi 3 generator to a new Fermi 3 switchyard. This new onsite transmission line corridor would be approximately 170 ft wide and include two sets of towers that would carry both rerouted Fermi 2 transmission lines and new Fermi 3 transmission lines (Detroit Edison 2011a). Surface water and wetland features located along the proposed onsite corridor include the south canal, a drainage area that is composed of a mosaic of emergent wetlands, and some forested wetlands (Detroit Edison 2011a). There are no surface water features within the footprint for the new switchyard (Detroit Edison 2011a). Clearing of the onsite transmission line ROW, erecting the transmission towers, and stringing of the transmission lines will all be accomplished using methods that minimize impacts on wetlands (Detroit Edison 2011a). The south canal and the drainage area within this portion of the Fermi site will be spanned by the transmission lines; impacts on the drainage area are expected to be minor because no activities associated with the transmission structure installation are expected to occur within the drainage channel (Detroit Edison 2011a).

Three new 345-kV transmission lines for Fermi 3 will be located within an assumed 300-ft-wide corridor from the Fermi site to the Milan Substation, with a length of approximately 29.4 mi. While the onsite Fermi 3 transmission lines will be owned by Detroit Edison up to the point of their interconnection with the new Fermi 3 switchyard, ITC*Transmission* will exclusively own and operate the offsite lines and other transmission system equipment between the Fermi 3 switchyard and the Milan Substation, and Detroit Edison will not control the building or operation of the transmission system. It is expected that Detroit Edison would contract with ITC*Transmission* to maintain the transmission towers and lines located on Detroit Edison property (Detroit Edison 2011a).

The transmission line corridor route is described in Section 2.4.1.2 of the EIS and is illustrated in Figure 2-5. The three 345-kV lines for Fermi 3 would be built in an east–west common corridor that currently contains transmission lines for Fermi 2 for approximately 5 mi to a point just west of I-75. From this point, the three Fermi-Milan lines would be in a corridor shared with other non-Fermi lines that travel to the west and north for approximately 13 mi. The last 10.8 mi of the proposed corridor that would proceed west to the Milan Substation are currently undeveloped, and no transmission infrastructure exists. This portion of the corridor has been under ITC*Transmission*'s control for future transmission development, but vegetation maintenance has been minimal except to remove tall, woody vegetation. According to FWS's National Wetland Inventory mapping, the identified transmission route crosses about 30 wetlands or other waters that may be regulated by the USACE and MDEQ (FWS 2010). The 18.6-mi existing eastern section of the transmission route crosses 12 narrow agricultural drains and small streams; the undeveloped western 10.8-mi section of the route crosses nine drains and small streams.

Impacts of transmission line development on aquatic resources along the eastern 18.6 mi of the transmission line corridor are expected to be small, since the reconfiguration of existing conductors would, for the most part, allow for the use of existing infrastructure (e.g., transmission line towers) to create the new lines, and access for installing additional lines is good because the vegetation has been managed to exclude tall woody vegetation. Existing aquatic habitats in this portion of the corridor would be spanned, and BMPs would be used to protect aquatic habitats crossed by the new lines. This includes, but is not limited to, the use of silt fencing and hay bales, and similar practices to ensure the protection of aquatic habitats in close proximity to building activity. Similarly, agricultural drains and small streams occurring in the undeveloped western corridor are narrow, and it is anticipated that placement of structures within stream channels could be avoided by using tower spans of 700–900 ft (Detroit Edison 2011a). Roads in the vicinity are expected to provide sufficient access to this region of the corridor without the need for building new access roads. There are no aquatic habitats within the area that would be affected by the anticipated expansion of the Milan Substation. Impacts of transmission line development on aquatic habitats within the proposed transmission line corridor would be temporary, easily mitigated, and minor, and no additional mitigation would be necessary.

4.0 Species Descriptions

This section identifies terrestrial and aquatic Federally listed species, including species that are candidates for Federal listing, that may occur on or near the Fermi site or the proposed transmission line corridors (see Table 1-1) and describes their life history and habitat use.

4.1 Terrestrial Species

4.1.1 Indiana Bat (*Myotis sodalis*)

The Indiana bat (*Myotis sodalis*) is Federally listed and State-listed as endangered. In its scoping letter, the FWS (2009a) identified the Indiana bat as potentially occurring in Monroe, Washtenaw, and Wayne counties, Michigan. The MDNR expressed no specific concern for the species during consultations in 2007 (Detroit Edison 2009b), and according to MNFI there are no reported occurrences of the Indiana bat in Monroe County (MNFI 2007b). No bats of any species were observed at the Fermi site during any of the wildlife surveys conducted by Detroit Edison since 2006. However, mist-net surveys for Indiana bats that follow FWS protocols have not been conducted on the Fermi site. MNFI records indicate that the Indiana bat has been observed in counties to the north and west of Monroe County. The species is found in Michigan only during late spring to early fall, when it roosts in forested areas beneath loose bark of large trees or in hollow snags (MNFI 2007b). With the death of many green ash (*Fraxinus pennsylvanica*) trees in the project area caused by the emerald ash borer (*Agrilus planipennis*), there are several trees that, at the time of the preparation of this biological assessment, may be suitable for summer roosting habitat (Detroit Edison 2011c).

On August 2, 2011, Detroit Edison conducted a field visit to the Fermi site to evaluate areas that would be affected by building the proposed Fermi 3 facilities and had not been investigated during the 2008–2009 survey because of site layout changes that occurred after the survey was completed (Detroit Edison 2011c). Detroit Edison evaluated potential roost trees in each location as low, moderate, or high potential based on criteria drawn from the FWS's *Indiana Bat (Myotis sodalis) Draft Recovery Plan: First Revision* (FWS 2007). Six trees were evaluated as potentially suitable for summer roosts by the Indiana bat and their locations were determined using a handheld GPS unit. Figure 4-1 illustrates these potential roost tree locations. Figure 4-1 also shows the location of a single large shagbark hickory tree identified during the 2008–2009 wildlife surveys in the woods east of Quarry Lakes Road.

One location was considered high potential, but this determination was based on a single tree that may deteriorate and become unsuitable by the time building of the transmission line

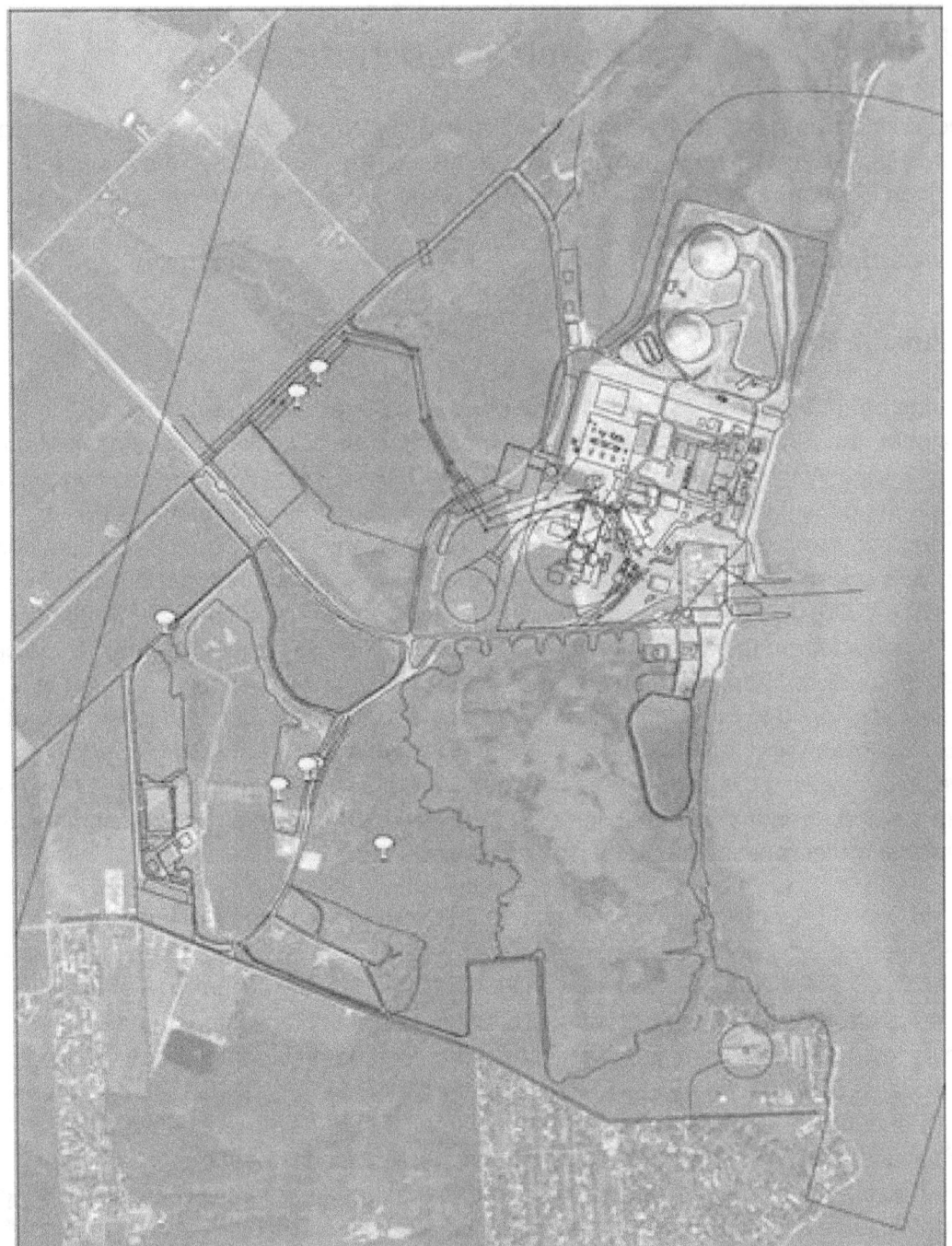

Figure 4-1. Fermi 3 Potential Indiana Bat Roost Trees (Source: Detroit Edison 2011c)

would occur. Most trees in the area were too small or otherwise unsuitable for Indiana bat summer roosts, and this situation is unlikely to improve with time.

Although some roost trees suitable for summer roosting habitat are present, other habitat features usually preferred by Indiana bat are generally lacking at the Fermi site. In addition, most of the potential roost trees are dead ash trees that will continue to deteriorate, so conditions for summer roosts will not improve before building of the new Fermi unit starts.

Because the transmission line corridor has not been surveyed, no potential roost trees have been identified in the corridor. It is possible that suitable roosting habitat occurs within the western 10.8-mi segment of the corridor. Other Indiana bat habitat features have likewise not been evaluated. Any ash trees in the corridor that have the potential to be potential roost trees at the time this BA was prepared would continue to deteriorate. Conditions for summer roosts would be unlikely to improve before the start of transmission line development.

4.1.2 Eastern Massasauga Rattlesnake (*Sistrurus catenatus catenatus*)

The eastern massasauga rattlesnake (*Sistrurus catenatus catenatus*) is a candidate for listing that is known or believed to occur in more than 50 counties in Michigan, including Washtenaw and Wayne counties (MNFI 2007a; FWS 2011b). This species is found in a variety of wetland habitats. Populations in southern Michigan are typically associated with open wetlands, particularly prairie fens. Some populations of the eastern massasauga rattlesnake also utilize open uplands and/or forest openings for foraging, basking, gestation, and parturition (i.e., giving birth to young) (Lee and Legge 2000; MNFI 2007a). Neither FWS nor MNFI have records of this snake occurring in Monroe County (MNFI 2007a; FWS 2011b). Therefore it is unlikely that the snake occurs on the Fermi site. No surveys have been conducted to evaluate the presence or absence of the snake or of suitable habitat along the transmission line corridor, including those areas where the corridor would pass through Washtenaw and Wayne counties.

4.1.3 Karner Blue Butterfly (*Lycaeides melissa samuelis*)

The Karner blue butterfly (*Lycaeides melissa samuelis*) is Federally listed as endangered, State-listed as threatened, and considered by FWS to occur in Monroe County but not on the Fermi site. It usually is associated with landscapes composed of sandy soils, which supported oak or oak-pine savanna or barrens prior to European settlement. Since its historical habitat suffers from fire suppression efforts, the butterfly often occurs in openings, old fields, and ROWs surrounded by close-canopied oak forest. Karner blue larvae feed exclusively on wild lupine (*Lupinus perennis*), but adults visit a wide variety of flowering plants for nectar (Rabe 2001). Although lupines were established in the prairie creation area in the onsite transmission line ROW and were observed in 2000 and 2002, no lupines were observed in subsequent vegetation surveys between 2006 and 2009 (Detroit Edison 2009a). This butterfly has not been observed in Washtenaw or Wayne counties, and it has not been seen in Monroe County since 1986 (MNFI 2007a). The most recent sitings of this butterfly have been in the west-central portion of lower Michigan (MNFI 2007a).

The MDNR Endangered Species Coordinator stated that Karner blue butterflies are not likely to occur on the Fermi site because none were found when the entire area was carefully surveyed in recent years prior to introduction of Karner blue butterflies in the Petersburg Wildlife Management Area near Petersburg, Michigan. The maximum movement of the butterflies from their point of introduction is about 1 km (Hoving 2010), eliminating the possibility that introduced butterflies would now occur on the Fermi site or along the transmission line corridor. In discussions between NRC and FWS in July 2011, FWS indicated that the Karner blue butterfly is unlikely to occur in the project area.

4.1.4 Mitchell's Satyr Butterfly (*Neonympha mitchellii mitchellii*)

The Mitchell's satyr butterfly (*Neonympha mitchellii mitchellii*) is Federally listed as endangered and is known or believed to occur in nine counties in southern Michigan, including Washtenaw County, but not Monroe or Wayne counties (FWS 2011c). It is also State listed as endangered (MNFI 2007a). Although its habitat requirements are not yet fully understood, this butterfly appears to be restricted to calcareous wetlands that range along a continuum from open fen, wet prairie, prairie fen, and sedge meadow to shrub-carr and tamarack savanna (Lee 2000; MNFI 2007a). According to the MNFI, this butterfly was last seen in Washtenaw County in 2010. According to the FWS, however, this species is unlikely to occur in the project area (Fisher 2011).

4.1.5 American Burying Beetle (*Nicrophorus americanus*)

The American burying beetle (*Nicrophorus americanus*) is Federally listed as endangered and State listed as presumed extirpated (meaning that the State believes that no individuals remain in the State). The species has not been observed on the Fermi site, and the last reported observation in the project area was in Washtenaw County in 1917 (MNFI 2007a). The American burying beetle formerly occupied a broad range of habitats, ranging from mature hardwood forests to old field shrubland to grassland. It is not found in sites with soils unsuitable to burying carrion, such as those with very loose sand, extremely dry soils, or saturated soils. The FWS did not mention this species in its scoping letter (FWS 2009a).

4.1.6 Eastern Prairie Fringed Orchid (*Platanthera leucophaea*)

The Eastern prairie fringed orchid (*Platanthera leucophaea*), also known as the prairie white-fringed orchid, is Federally listed as threatened and State listed as endangered. This species has not been observed on or near the Fermi site in any vegetation studies conducted on the site since 1973, but it has been reported in Monroe County as recently as 2006 (MNFI 2007c). The plant is known mostly from lakeplain prairies around Saginaw Bay and western Lake Erie, occurring in moist alkaline and lacustrine soils. This habitat is not present on the Fermi site or in the immediate vicinity, but it may occur along the proposed transmission line corridor. Although it is rare, this orchid can readily colonize highly disturbed sites such as ditches, unmowed old fields, and even the edges of golf courses, as long as competition is not overly intense and proper soil fungi are present. No surveys have been conducted to evaluate the presence or absence of this orchid along the transmission line corridor.

4.2 Aquatic Species

4.2.1 Northern Riffleshell (*Epioblasma torulosa rangiana*)

The northern riffleshell (*Epioblasma torulosa rangiana*) is a freshwater unionid mussel (see Section 2.4.2.1 of the EIS) that was Federally listed as an endangered species in 1993 (58 FR 5638) and is also listed as endangered by the State of Michigan (MNFI 2007a). The historic range for the northern riffleshell includes Illinois, Indiana, Kentucky, Michigan, Ohio, Pennsylvania, West Virginia, and western Ontario (Carman and Goforth 2000a). It was once widespread in the Ohio and Maumee River Basins and in tributaries of western Lake Erie (Carman and Goforth 2000a). In Michigan, the northern riffleshell is only known to currently occur in the Black River in Sanilac County and the Detroit River in Wayne County (Carman and Goforth 2000a). More than 100 individuals from the Detroit River population were relocated to the St. Clair River in 1992 as part of an effort to establish a new population, but the success of that effort is unknown (Carman and Goforth 2000a).

The habitat for the northern riffleshell is fine to coarse gravel in riffles and runs of streams with swift currents (MNFI 2007a). The northern riffleshell was last observed in Monroe County in 1977 and in Wayne County in 2006 (MNFI 2007a). The northern riffleshell has not been reported from Washtenaw County (MNFI 2007a). No streams with conditions suitable for the northern riffleshell are present on the Fermi site; whether appropriate habitats are present in stream areas that are crossed by the proposed transmission line corridor is currently unknown. No extant populations of this species are known from stream drainages that would be crossed by the transmission lines. The northern riffleshell is a riverine species and areas of Lake Erie adjacent to the Fermi site do not offer suitable habitat for this species.

4.2.2 Rayed Bean (*Villosa fabalis*)

The rayed bean (*Villosa fabalis*) is a freshwater unionid mussel that was Federally listed as an endangered species in 2012 (77 FR 8632). This species is also listed as endangered by the State of Michigan and has been recorded in Monroe and Wayne Counties (MNFI 2007a). The rayed bean is patchily distributed in the St. Lawrence, Ohio, and Tennessee River drainages (Carman 2001). Although it was historically widespread from Ontario to Alabama and Illinois to New York, only a few populations are currently known to exist, and it is assumed to be extirpated throughout much of its former range (Carman 2001). Extant populations are currently known from 31 streams in Indiana, Michigan, New York, Ohio, Pennsylvania, Tennessee, and West Virginia in the United States, and the province of Ontario in Canada (77 FR 8632). In Michigan, existing rayed bean populations are known from the Black, Pine, Belle, and Clinton River systems (77 FR 8632).

The rayed bean is generally found in smaller headwater creeks, although it has also been found in larger rivers (FWS 2002). It usually is found in or near shoal or riffle areas; there are also records of rayed bean specimens (valves only) from shallow, wave-washed areas of Lake Erie generally associated with islands in the western portion of the lake (FWS 2002). Preferred substrates are gravel and sand, and it is oftentimes found among the roots of vegetation growing in riffles and shoals (FWS 2002). The rayed bean has experienced a significant reduction in range, and most of its populations are isolated and appear to be declining (FWS 2002). The survival of the rayed bean is threatened by a variety of stressors, especially habitat destruction associated with siltation, dredging, and channelization, and the introduction of alien species such as the Asian clam and zebra and quagga mussels (FWS 2002).

Valves of the rayed bean were last observed in Monroe County in 1984 and in Wayne County in 2006 (MNFI 2007a). These observations were based upon the presence of shells, not living specimens (Carman 2001). The rayed bean has not been reported from Washtenaw County (MNFI 2007a). There are no streams on the Fermi site with conditions suitable for the rayed bean, and no extant populations are known to occur in the stream drainages that would be crossed by the proposed transmission line route.

Although there are records of rayed bean valves from shallow, wave-washed areas of western Lake Erie, information supplied by Detroit Edison suggests that it is unlikely that the species occurs in the vicinity of the Fermi site for a number of reasons: (1) approximately 30 yr of information on mussels in the western basin of Lake Erie (including in the vicinity of the Fermi site) have been collected and evaluated by the U.S. Geological Survey (USGS) and no rayed bean specimens have been identified; (2) the USACE conducted mussel surveys in Lake Erie approximately 2 mi south of the Fermi site and found no live specimens or shells of the rayed bean; (3) the rayed bean was not observed in surveys conducted by the MNFI just north of the Fermi site near the mouth of Swan Creek; and (4) observations made by divers during sediment sampling and buoy maintenance activities within the exclusion zone for the Fermi site indicate that the sediment is predominantly clay hardpan, which would not be suitable for the rayed bean (Detroit Edison 2010).

4.2.3 Snuffbox Mussel (*Epioblasma triquetra*)

The snuffbox mussel (*Epioblasma triquetra*) is a freshwater unionid mussel that was Federally listed as an endangered species in 2012 (77 FR 8632). This species is also listed as endangered by the State of Michigan and has been recorded in Monroe, Wayne, and Washtenaw Counties (MNFI 2007a). The historic range of the snuffbox mussel extends from Ontario southward to Mississippi and Alabama and eastward to New York and Virginia; extant populations are still present in Wisconsin, Illinois, Indiana, Kentucky, Michigan, Ohio, Pennsylvania, Tennessee, and West Virginia (NatureServe 2009). In Michigan, this species is found primarily in eastern and southeastern rivers and has been reported from Otter Creek in Monroe County and the Detroit River in Wayne County (Carman and Goforth 2000b). The snuffbox mussel primarily inhabits small and medium-sized rivers, although specimens have also been collected from Lake Erie and large rivers, such as the St. Clair River. Preferred habitat usually has clear water and sand, gravel, or cobble substrate with a swift current; individuals are often buried deep in the sediment (Carman and Goforth 2000b). In Michigan, the only known fish host is the log perch (*Percina caprodes*), although the banded sculpin (*Cottus carolinae*) has been identified as a fish host in other portions of its range (Carman and Goforth 2000b).

The snuffbox mussel was last reported from Monroe, Wayne, and Washtenaw Counties in 1933, 2000, and 1977, respectively (MNFI 2007a). Streams with conditions suitable for the snuffbox mussel are not present on the Fermi site. Although there is a possibility that shoreline areas of Lake Erie near the Fermi site could contain suitable substrates for the snuffbox mussel, information supplied by Detroit Edison suggests that it is unlikely that the species occurs in the vicinity of the Fermi site for a number of reasons: (1) approximately 30 yr of information on mussels in the western basin of Lake Erie (including in the vicinity of the Fermi site) have been collected and evaluated by the USGS and no snuffbox mussel specimens have been identified; (2) the USACE conducted mussel surveys in Lake Erie approximately 2 mi south of the Fermi site and found no live specimens or shells of the snuffbox mussel; (3) the snuffbox mussel was not observed in surveys conducted by the MNFI just north of the Fermi site near the mouth of Swan Creek; and (4) observations made by divers during sediment sampling and buoy maintenance activities within the exclusion zone for the Fermi site indicate that the sediment is predominantly clay hardpan, which would not be suitable for the snuffbox mussel (Detroit Edison 2010).

It is currently unknown whether appropriate habitats for the snuffbox mussel are present in any of the streams that are crossed by the proposed transmission line corridor. However, no extant populations of this species are known from stream drainages that would be crossed by the transmission lines.

5.0 Potential Environmental Effects of the Proposed Actions

This section describes the potential impacts from building and operating the proposed Fermi 3 on the species listed in Table 1-1.

5.1 Building Impacts

The following paragraphs describe the potential for building of Fermi 3 to affect Federally listed species, including species that are candidates for Federal listing, with the potential to occur on and within the vicinity of the Fermi site and transmission line corridors (see Table 1-1).

5.1.1 Fermi Site

5.1.1.1 Terrestrial Species

Indiana Bat

The Indiana bat has not been observed on the Fermi site, nor has it been reported in Monroe County by the MNFI (MNFI 2007b); however, the Indiana bat has been observed in counties to the north and west of Monroe County (MNFI 2007a). The Indiana bat is known to or believed to occur in Monroe County, according to the FWS. There is currently a low probability that suitable habitat for this species might be present on the Fermi site; that probability will decrease in the next few years as the dead and dying ash trees further deteriorate. Considering there is a low probability that suitable Indiana bat habitat would exist on the Fermi site by the time building activities would begin, the review team concludes that the proposed site work may affect, but would not be likely to adversely affect, the Indiana bat.

Eastern Massasauga Rattlesnake

There is no record of occurrence of this species in Monroe County (MNFI 2007a; FWS 2011b); thus, it is unlikely to occur on the Fermi site. Therefore, the review team concludes that project-related building activities at the Fermi site would have no effect on the eastern massasauga rattlesnake.

Karner Blue Butterfly

Although the Karner blue butterfly was observed in Monroe County in 1986 (MNFI 2007a), the FWS stated that this species is unlikely to occur in the project area (FWS 2011a). MDNR also stated that the probability of the Karner blue butterfly occurring on the Fermi site is very low. Therefore, the review team concludes that project-related building activities at the Fermi site would have no effect on the Karner blue butterfly.

Mitchell's Satyr Butterfly

Although MNFI records indicate the insect was observed in Washtenaw County in 2010, according to the FWS, this species is unlikely to occur in the project area (Fisher 2011). Therefore, project-related building activities at the Fermi site would have no effect on the Mitchell's satyr butterfly.

American Burying Beetle

This species is presumed extirpated from the State and has not been seen in the project area since 1917 (MNFI 2007a). Because it is unlikely to occur anywhere in the State, project-related building activities at the Fermi site would have no effect on the American burying beetle.

Eastern Prairie Fringed Orchid

There are no recorded occurrences of this species on the Fermi site (MNFI 2007a; FWS 2011a), and it was not observed during any of the plant or wildlife surveys conducted on the Fermi site (Detroit Edison 2009a). The FWS has indicated the plant is unlikely to occur on the Fermi site (Fisher 2011). However, this species was observed in Monroe County within the last 5 years, and the plant is known to occur in lakeplain prairies around western Lake Erie. The plant may therefore occur on the Fermi site. There are approximately 238 ac of emergent wetlands on the Fermi site. Because Fermi 3 would impact only approximately 20.9 ac (about 9 percent) of the emergent wetlands on the Fermi site and because large portions of these emergent wetlands are likely to be unsuitable because they have been severely degraded by the common reed, an invasive plant, the review team has determined that project-related building activities on the Fermi site may affect, but would be unlikely to adversely affect, the Eastern prairie fringed orchid.

5.1.1.2 Aquatic Species

Northern Riffleshell

There is no suitable habitat for the northern riffleshell on the Fermi site or in adjacent waters of Lake Erie (Section 4.2). In addition, there are no recent records of occurrence of this species from the Fermi site or environs. On the basis of this information, the review team concludes that project-related building activities on the Fermi site would have no effect on the northern riffleshell.

Rayed Bean

There are no streams on the Fermi site with conditions suitable for the rayed bean. Although there are records of rayed bean valves from shallow, wave-washed areas of western Lake Erie, it is considered unlikely for the species to occur in the vicinity of the Fermi site for a number of reasons, as presented in Section 4. In addition, most of the area that would be affected by development of the intake structure, barge slip, and the discharge structure for Fermi 3 has been disturbed previously by periodic maintenance dredging. On the basis of this information, the review team concludes that project-related building activities on the Fermi site would have no effect on the rayed bean.

Snuffbox Mussel

There are no recent records of occurrence of this species from the Fermi site or environs. Although there are no suitable stream habitats on the Fermi site, there is the potential for suitable habitats to exist in Lake Erie and the host required by this species (logperch, *Percina caprodes*) has been collected near the Fermi site in Swan Creek and in Lake Erie near the South Lagoon (AECOM 2009). The areas in Lake Erie that would be disturbed during the building of Fermi 3 facilities either have been previously disturbed by periodic maintenance dredging or have a clay hardpan substrate (Detroit Edison 2010) rather than the sand, gravel, or cobble substrate preferred by this species. Therefore, it is considered unlikely that this species would be present in the project area. On the basis of this information, the review team concludes that project-related building activities on the Fermi site would have no effect on the snuffbox mussel.

5.1.2 Transmission Line Corridors

Although ITC*Transmission* has not finalized the locations of the transmission line corridor or ancillary areas (e.g., laydown areas), Detroit Edison has indicated that the proposed route is the result of past analyses of routes conducted for the development of Fermi 2. The route analysis (Detroit Edison 2009b) followed guidance from the U.S. Department of the Interior and the Federal Power Commission for siting transmission lines. In addition, other criteria were considered to minimize environmental impacts (Detroit Edison 2011a).

5.1.2.1 Terrestrial Species

Because ITC*Transmission* has not yet performed on-the-ground field surveys for Federally listed species along the proposed routes, the review team consulted online sources, including the MNFI and the FWS Environmental Conservation Online System, to determine what information is currently available. Once final routes have been determined, ITC*Transmission* is expected to conduct on-the-ground field surveys for each line prior to completing applications for the required USACE and MDNR permits. ITC*Transmission* would likely have to implement BMPs to minimize any potential impacts on Federally listed species and critical habitats during transmission line development activities, based on USACE and MDNR permit conditions (Detroit Edison 2011a).

The Milan substation is expected to be expanded from its current dimensions of 300 ft by 500 ft to approximately 1000 ft by 1000 ft. All of the area that would be used for the expansion is either cropland or mowed grass. Building an expanded substation would, therefore, have no effect on any of the Federally listed terrestrial species, including species that are candidates for Federal listing.

Indiana Bat

The Indiana bat has been observed in Washtenaw and Wayne counties (MNFI 2007a), and this species might occur in suitable habitat along the transmission line corridor. The review team believes that if Detroit Edison limits clearing of forest cover to between October 15 and March 31, it could avoid potentially adverse effects on Indiana bats (FWS 2009b). Detroit Edison and ITC *Transmission* could also avoid adverse effects by conducting surveys of suitable habitat trees similar to the survey conducted on the Fermi site in August 2011. If the results of such a survey failed to locate suitable habitat, the likelihood of Indiana bats being present and affected by building the transmission line would be minimal. Alternately, Detroit Edison or ITC *Transmission* could conduct targeted mist nest surveys of forested areas using an FWS-approved protocol (FWS 2011d) prior to disturbance and proceed only if the surveys reveal that no bats are present. Considering the ability to avoid adverse impacts, the review team concludes that building of the proposed transmission lines may affect, but is not likely to adversely affect, the Indiana bat.

Eastern Massasauga Rattlesnake

Because there is no record of occurrence of this species in Monroe County, it is unlikely to occur along the Monroe County segment of the transmission line route. Based on its known distribution, this snake could occur in wetlands and naturally vegetated upland habitats crossed by the Washtenaw County and Wayne County segments of the route. Clearing forested wetlands would be necessary to establish new transmission corridor. Forested wetlands within the transmission line ROW would be converted to emergent or scrub-shrub wetlands for the long term. Because the species favors both open and forested wetlands (MNFI 2007a), conversion of wetland from forested to emergent or scrub-shrub wetland is unlikely to adversely affect the species. The same is true for clearing upland forests for the transmission lines.

The greatest potential for impacts on this snake would be during ground disturbance of naturally vegetated areas to build tower pads and access roads. The eastern massasauga rattlesnake is a mobile snake, and active adults with a length of 2 to 3 ft would likely move out of the way before being crushed by construction equipment. However, ground disturbance of nests or underground hibernation areas could kill or injure individuals. ITC *Transmission* could reduce the potential for impacts by surveying areas subject to ground disturbance prior to clearing and grubbing, delaying work if hibernation areas or nests are discovered, and relocating discovered individuals to nearby naturally vegetated areas. Recognizing the possibility of these simple management efforts, the review team concludes that building of the proposed transmission lines may affect, but is not likely to adversely affect, the eastern massasauga rattlesnake.

Karner Blue Butterfly

The Karner blue butterfly is listed as endangered and is recognized as potentially occurring in Monroe County (MNFI 2007a). It has not been seen in Monroe County since 1986 (MNFI 2007a; FWS 2011a). This butterfly has not been observed in Washtenaw or Wayne counties (MNFI 2007a) and is unlikely to occur in the project area in those counties. The review team therefore concludes that building of the proposed transmission line would have no effect on the Karner blue butterfly.

Mitchell's Satyr Butterfly

Although MNFI records indicate this insect was observed in Washtenaw County in 2010, according to the FWS, this species is unlikely to occur in the project area (Fisher 2011). Therefore, the review team concludes that building the proposed transmission lines would not affect the Mitchell's satyr butterfly.

American Burying Beetle

This species is presumed extirpated from the State and has not been seen in the project area since 1917 (MNFI 2007a). Because it is unlikely to occur anywhere in the State, building the proposed transmission lines would not affect the American burying beetle.

Eastern Prairie Fringed Orchid

The Eastern prairie fringed orchid has been observed in recent years in Monroe, Washtenaw, and Wayne counties. This plant could potentially occur wherever suitable habitat exists along the proposed transmission line route. Because the plant favors open rather than forested wetland habitat, forest clearing to establish new transmission line corridor lands is unlikely to result in adverse effects. However, filling emergent wetlands to build tower pads or access roads could kill any specimens present within the filled wetlands. However, it should be possible for ITC*Transmission* to survey emergent wetlands for this plant prior to any fill activities and make minor adjustments to tower placements or access road alignments to avoid any identified specimens. Recognizing that ITC*Transmission* could use BMPs or make minor adjustments to wetland locations affected by building of the transmission line, the review team concludes that building the proposed transmission lines may affect, but is not likely to adversely affect, the Eastern prairie fringed orchid.

5.1.2.2   Aquatic Species

Northern Riffleshell

Although suitable habitat for the northern riffleshell could be present in some of the streams that would be crossed by the proposed transmission line corridor, it is not expected to occur along the transmission line route because extant populations of this species in Michigan are only known to be present in the Black River in Sanilac County and the Detroit River in Wayne County (Carman and Goforth 2000a). Even if present in streams crossed by the transmission line corridors, building transmission lines for Fermi 3 is not expected to affect the northern riffleshell because aquatic habitats that are crossed by the corridor would be spanned without placement of structures within stream channels and because BMPs would be implemented to protect water quality in aquatic habitats located near building activity. Additional regulatory review of proposed plans for building of the transmission lines, which would be built, owned, and maintained by ITC*Transmission*, would be conducted by MDNR, and potential impacts on water quality are expected to be addressed through mitigation measures and BMPs required under other State- or Federally issued permits. On the basis of this information, the review team concludes that building of transmission lines for Fermi 3 would have no effect on the northern riffleshell.

Rayed Bean

No extant populations are known to occur in the stream drainages that would be crossed by the proposed transmission line route. The building of transmission lines for Fermi 3 is not expected to affect the rayed bean because the species has not been reported from the streams that would be crossed by the proposed transmission line corridor; aquatic habitats that are crossed by the corridor would be spanned without placement of structures within stream channels; and BMPs would be implemented to protect water quality in aquatic habitats located near building activity. On the basis of this information, the review team concludes that building transmission lines for Fermi 3 would have no effect on the rayed bean.

Snuffbox Mussel

It is not known whether suitable stream habitat or populations of the snuffbox mussel occur along the proposed offsite transmission line corridor. However, no extant populations of this species are known from stream drainages that would be crossed by the transmission lines. It is anticipated that the small streams that would be crossed by the proposed transmission line corridor could be easily spanned without placing structures in stream channels and that BMPs would be implemented to protect water quality in streams during building activities. Additional regulatory review of proposed plans for building the offsite transmission lines, which would be built, owned, and maintained by ITC*Transmission*, would be conducted by MDNR, and potential impacts on water quality are expected to be addressed through mitigation measures and BMPs required under other State- or Federally issued permits. On the basis of this information, the review team concludes that building the transmission lines for Fermi 3 would have no effect on the snuffbox mussel.

5.2 Operations Impacts

The following paragraphs describe the potential for operations-related impacts on the Federally listed species, including species that are proposed or candidates for Federal listing with the potential to occur on and within the vicinity of the Fermi site and transmission line corridors (see Table 1-1).

5.2.1 Fermi Site

5.2.1.1 Terrestrial Species

Indiana Bat

This species has potential to occur in suitable habitat on the Fermi site. This species might roost and forage in forested and other naturally vegetated suitable habitats on the Fermi site. However, those habitats would not be disturbed by Fermi 3 operations. Therefore, the operation of Fermi 3 would have no effect on the Indiana bat on the Fermi site.

Eastern Massasauga Rattlesnake

There is no record of occurrence of this species in Monroe County (MNFI 2007a; FWS 2011b); thus, the snake is unlikely to occur on the Fermi site. Therefore, operation of Fermi 3 would have no effect on the eastern massasauga rattlesnake on the Fermi site.

Karner Blue Butterfly

The Karner blue butterfly has not been seen in Monroe County since 1986 (MNFI 2007a), and the FWS stated that this species is unlikely to occur in the project area (FWS 2011a). MDNR also stated that the probability of the Karner blue butterfly occurring on the Fermi site is very low. Therefore, the review team concludes that the operation of Fermi 3 would have no effect on the Karner blue butterfly on the Fermi site.

Mitchell's Satyr Butterfly

Although MNFI records indicate this insect was observed in Washtenaw County in 2010, according to the FWS, this species is unlikely to occur in the project area (Fisher 2011). Therefore, the review team concludes that the operation of Fermi 3 would have no effect on the Mitchell's satyr butterfly.

American Burying Beetle

This species is presumed extirpated from the State and has not been seen in the project area since 1917 (MNFI 2007a). The FWS did not mention this species in its scoping letter (FWS 2009a). Therefore, operation of Fermi 3 would have no effect on this insect on the Fermi site.

Eastern Prairie Fringed Orchid

There are no recorded occurrences of this species on or near the Fermi site (MNFI 2007a; FWS 2011a). However, the plant is known mostly from lakeplain prairies around Saginaw Bay and western Lake Erie; therefore, this plant may occur on the Fermi 3 site. Nevertheless, even if specimens occurred in wetland habitats on the site, operations would not disturb wetland habitats. Therefore, the operation of Fermi 3 would have no effect on the Eastern prairie fringed orchid on the Fermi site.

5.2.1.2 Aquatic Species

Northern Riffleshell

There are no recent records of occurrence of this species from the Fermi site or environs. As identified in Section 2.2, there are no streams on the Fermi site with conditions suitable for the northern riffleshell. In addition, the northern riffleshell is a riverine species that would not occur in Lake Erie (Section 2.2). Because there is no suitable habitat for the northern riffleshell on the Fermi site or in adjacent waters of Lake Erie, the operation of Fermi 3, including withdrawal and discharge of cooling water from or into Lake Erie and NPDES-permitted discharges waste water and storm water into onsite water bodies, would have no effect on this species. Further, it is anticipated that water quality would be maintained during operations because (1) the NPDES permit for Fermi 3 would specify allowable concentrations of chemicals in Fermi 3 discharges and would require regular testing to evaluate compliance, and (2) Detroit Edison has stated that the Fermi 3 SWPPP and design features would be used to control stormwater runoff to ensure that sediment loading to Swan Creek is adequately controlled to minimize water quality impacts (Detroit Edison 2011a). On the basis of this information, the review team concludes that operation of Fermi 3 would have no effect on the northern riffleshell.

Rayed Bean

As identified in Section 2.2, there are no streams on the Fermi site with conditions suitable for the rayed bean and it is believed that the species is unlikely to be present in Lake Erie near the Fermi site. Even if the rayed bean was present in the vicinity of the Fermi site, periodic dredging would be unlikely to affect the species within the project area, because the intake bay has been dredged in the past. As a consequence, it is unlikely that the substrate within areas that would periodically require dredging during Fermi 3 operations would be suitable for the rayed bean.

As eggs, unionid mussels are not likely to be affected by entrainment through the cooling water intake because they are not free-floating, but rather develop into larvae within the female. The glochidial stage, during which juvenile mussels attach to a suitable fish host, may be indirectly vulnerable through impingement and entrainment of host species. Post-glochidial and adult stages are not likely to be susceptible to entrainment because they bury themselves in sediment. Fish hosts for the glochidia of the rayed bean could include the Tippecanoe darter (*Etheostoma tippecanoe*), greenside darter (*Etheostoma blennioides*), rainbow darter (*Etheostoma caeruleum*), mottled sculpin (*Cottus bairdi*), and largemouth bass (*Micropterus salmoides*). Of these potential host species, only the largemouth bass was observed in fish collections in Lake Erie near the intake structure or near the discharge from the South Lagoon. Based on impingement studies conducted at the existing Fermi 2 intake in 2008 and 2009, it is estimated that small numbers of largemouth bass individuals (approximately 30) would be impinged annually with the intake pumps for Fermi 3 at full operating capacity (AECOM 2009).

It is anticipated that operation of Fermi 3 would not result in water quality unsuitable for the rayed bean if a population were present in Lake Erie near the Fermi site. Thermal effects associated with cooling water discharge during operation of Fermi 3 would be unlikely to affect mussels, as the discharge ports would direct water upward and not toward the lake bottom. In addition, it is anticipated that suitable water quality would be maintained because (1) the NPDES permit for Fermi 3 would specify allowable concentrations of chemicals in the Fermi 3 discharge and would require regular testing to evaluate compliance, and (2) Detroit Edison has stated that the Fermi 3 SWPPP and design features would be used to control stormwater runoff to ensure that sediment loading to Swan Creek and/or Lake Erie is adequately controlled to minimize water quality impacts (Detroit Edison 2011a). On the basis of the above information, the review team concludes that of the operation of Fermi 3 would have no effect on the rayed bean.

Snuffbox Mussel

Although there are no suitable stream habitats on the Fermi site, there is potential for suitable habitats in adjacent areas of Lake Erie and the host required by this species (logperch, *Percina caprodes*) has been collected from the Fermi site at sampling locations in Swan Creek and in

Lake Erie near the South Lagoon. Even if the snuffbox mussel was present in the vicinity of the Fermi site, periodic dredging would be unlikely to affect the species within the project area, because the intake bay has been dredged in the past. As a consequence, it is unlikely that the substrate within areas that would periodically require dredging during Fermi 3 operations would be suitable for the snuffbox mussel.

As eggs, unionid mussels are not likely to be affected by entrainment through the cooling water intake because they are not free-floating, but rather develop into larvae within the female. The glochidial stage, during which juvenile mussels attach to a suitable fish host, may be indirectly vulnerable through impingement and entrainment of host species. Post-glochidial and adult stages would not be susceptible to entrainment because they bury themselves in sediment. Fish hosts for the snuffbox mussel include the logperch, which was observed in fish collections in Lake Erie near the discharge from the South Lagoon and in Swan Creek. Based on impingement studies conducted during 1991 and 1992, Lawler, Matusky, and Skelly Engineers (1993) estimated that approximately 31 logperch were impinged annually by the Fermi 2 cooling water intake. However, impingement studies conducted during 2008 and 2009 at the Fermi 2 intake did not observe impingement of any logperch (AECOM 2009). Together, these two impingement studies suggest that small numbers of logperch could be impinged by the operation of the cooling water intake for Fermi 3.

It is anticipated that operation of Fermi 3 would not result in water quality unsuitable for the snuffbox mussel if a population were present in Lake Erie near the Fermi site. Thermal effects associated with cooling water discharge during operation of Fermi 3 would be unlikely to affect mussels, as the discharge ports would direct water upward and not toward the lake bottom. In addition, it is anticipated that suitable water quality would be maintained because (1) the NPDES permit for Fermi 3 would specify allowable concentrations of chemicals in the Fermi 3 discharge and would require regular testing to evaluate compliance, and (2) Detroit Edison has stated that the Fermi 3 SWPPP and design features would be used to control stormwater runoff to ensure that sediment loading to Swan Creek and/or Lake Erie is adequately controlled to minimize water quality impacts (Detroit Edison 2011a). On the basis of the above information, the review team concludes that the operation of Fermi 3 would have no effect on the snuffbox mussel.

5.2.2 Transmission Line Corridors

5.2.2.1 Terrestrial Species

Indiana Bat

This species has potential to occur in suitable habitat along the transmission line corridor. This species might roost and forage in forested and other naturally vegetated suitable habitats on the

Fermi site. However, those habitats would not be disturbed by operation of the Fermi 3 transmission lines. Therefore, the review team concludes that the operation of the Fermi 3 transmission lines would have no effect on the Indiana bat.

Eastern Massasauga Rattlesnake

This species may occur in Washtenaw and Wayne counties along the transmission line corridor. However, as discussed above, the eastern massasauga rattlesnake is a mobile snake and active adults with a length of 2 to 3 ft would likely move to avoid temporary impacts associated with transmission line corridor maintenance. Consequently, if Detroit Edison and ITC*Transmission* (1) conduct surveys to identify whether the eastern massasauga rattlesnake or its habitat occur along or adjacent to the proposed transmission line corridors, (2) are flexible in routing to avoid such sites, (3) implement BMPs to minimize impacts, and (4) adhere to Federal and State laws, the review team concludes that operation of the Fermi 3 transmission lines would have no effect on the eastern massasauga rattlesnake.

Karner Blue Butterfly

The Karner blue butterfly is listed as endangered and is recognized as potentially occurring in Monroe County (MNFI 2007a). It has not been seen in Monroe County since 1986 (MNFI 2007a; FWS 2011a). This butterfly has not been observed in Washtenaw or Wayne counties (MNFI 2007a) and is unlikely to occur along the proposed route. The review team therefore concludes that operation of the transmission line would have no effect on the Karner blue butterfly.

Mitchell's Satyr Butterfly

Although MNFI records indicate this insect was observed in Washtenaw County in 2010, according to the FWS, this species is unlikely to occur in the project area (Fisher 2011). Therefore, the review team concludes operation of the Fermi 3 transmission lines would not affect the Mitchell's satyr butterfly.

American Burying Beetle

The last reported observation of this species in the project area was in Washtenaw County in 1917 (MNFI 2007a). The State status of this insect is presumed extirpated. The FWS did not mention this species in its scoping letter (FWS 2009a). Therefore, the review team concludes operation of the Fermi 3 transmission lines would have no effect on the American burying beetle.

Eastern Prairie Fringed Orchid

The Eastern prairie fringed orchid could potentially occur wherever suitable habitat exists along the proposed transmission line route. Therefore, the review team has determined that operation of the proposed project may affect the Eastern prairie fringed orchid in the proposed transmission line corridors. However, if Detroit Edison and ITC *Transmission* (1) conduct surveys to identify whether the Eastern prairie fringed orchid occurs along or adjacent to the proposed transmission line corridors, (2) are flexible in routing to avoid such sites, (3) implement BMPs to minimize impacts associated with vegetation control activities, and (4) adhere to Federal and State laws, the review team concludes operation of the Fermi 3 transmission lines may affect, but would not likely adversely affect, the Eastern prairie fringed orchid.

5.2.2.2   Aquatic Species

Northern Riffleshell

Although suitable habitat for the northern riffleshell could be present in some of the streams that would be crossed by the proposed transmission line corridor, the species is not expected to occur along the transmission line route because extant populations in Michigan are only known to be present in the Black River in Sanilac County and the Detroit River in Wayne County (Carman and Goforth 2000a). Even if present in streams crossed by the transmission line corridors, impacts on the northern riffleshell from maintenance of transmission lines are unlikely, provided that BMPs identified in permits for the transmission lines are implemented. Additional regulatory review and permitting of proposed plans for maintenance of the transmission lines (e.g., for annual vegetation management plans) would be required prior to implementation (Detroit Edison 2011a). On the basis of this information, the review team concludes that operation and maintenance of transmission lines for Fermi 3 would have no effect on the northern riffleshell.

Rayed Bean

No extant populations of the rayed bean are known to occur in the stream drainages that would be crossed by the proposed transmission line route. The operation and maintenance of transmission lines for Fermi 3 are not expected to affect the rayed bean because the species has not been reported from the streams that would be crossed by the proposed transmission line corridor, because structures requiring maintenance (e.g., transmission towers) would not be placed in aquatic habitats that are crossed by the corridor, and because BMPs would be implemented to protect water quality in aquatic habitats during maintenance activities such as vegetation management (Detroit Edison 2011a). On the basis of this information, the review team concludes that operation and maintenance of transmission lines for Fermi 3 would have no effect on the rayed bean.

Snuffbox Mussel

It is not known whether suitable stream habitats for, or populations of, the snuffbox mussel occur along the proposed transmission line corridor. However, no extant populations of this species are known from stream drainages that would be crossed by the transmission lines. Even if present, impacts on the snuffbox mussel from the operation and maintenance of transmission lines for Fermi 3 are not anticipated because structures requiring maintenance (e.g., transmission towers) would not be placed in aquatic habitats that are crossed by the corridor, and BMPs would be implemented to protect water quality in aquatic habitats during maintenance activities such as vegetation management (Detroit Edison 2011a). On the basis of this information, the review team concludes that operation and maintenance of transmission lines for Fermi 3 would have no effect on the snuffbox mussel.

6.0 Cumulative Effects

6.1 Terrestrial

Current projects within the geographic area of interest potentially capable of affecting the same terrestrial ecological resources as Fermi 3, including the new transmission lines, include the ongoing operation of Fermi 2, the Detroit Edison Monroe Power Plant, the Bayshore Power Plant, the J.R. Whiting Power Plant, three limestone quarries, and several wastewater treatment plants (see Table 7-1 in the Fermi 3 Draft EIS). Reasonably foreseeable projects within the geographic area of interest that could affect the same terrestrial ecological resources include expanded regional commercial and residential development, building of the Ventower Industries manufacturing facility, and building of a proposed Cleveland-Toledo-Detroit passenger rail line. Ongoing commercial and residential development in the region would be expected to add to the loss of various habitats and wildlife, but the review team has no information about specific individual development proposals.

The geographic area of interest includes agricultural land, including row crops; open water, including part of Lake Erie and shallow lagoons within the Fermi site; developed land, especially in the Detroit metropolitan area; upland forests; and forested and emergent wetlands. With the exception of Great Lakes marsh and southern hardwood swamp, the habitats and wildlife that would be disturbed are common in the region. The habitats that would be affected as a result of any of the reasonably foreseeable activities listed above are not considered unique or critical for the survival of Federally listed threatened or endangered species or for the other important species identified in Section 2.4.1 of the Fermi 3 Draft EIS.

In the vicinity of the proposed transmission line corridors, potential future activities that could contribute to effects on threatened and endangered terrestrial species also include the potential expansion of the existing transmission system and the potential for other development activities, both residential and commercial, in the vicinity of the proposed transmission line corridors. Whether such activities or development will occur and the level of development that could be realized are unknown. However, such development could result in further loss of habitat and increased forest fragmentation that could affect species that inhabit those areas.

At least some of the other current and potential projects as listed above in the area of interest would affect some of the same habitats as the Fermi 3 project. It can therefore be concluded that one or more of them may also affect some of the same Federally listed species that are identified in Table 1-1. However, the habitats that would be affected are not considered unique or critical for the survival of Federally listed terrestrial species, including species that are candidates for Federal listing. As described in Section 5 of this BA, the Fermi 3 project may affect, but is not likely to adversely affect, Federally listed terrestrial species. None of the available information concerning other projects in the site vicinity suggest a potential for habitat disruption, or other environmental effects that would cause a noticeable impact to the terrestrial species when combined with building and operating a new ESBWR unit at the Fermi site. Therefore, the review team concludes that the contribution of building and operating Fermi 3 to the cumulative impacts on Federally listed terrestrial species is likely to be minimal.

6.2 Aquatic

In addition to the impacts from building and operation of Fermi 3 and the associated transmission facilities, the cumulative analysis considers other past, present, and reasonably foreseeable future actions that could affect aquatic resources within the same watersheds that could be affected by building and operation of Fermi 3. The geographic area of interest for the cumulative impact analysis for aquatic resources includes primarily the lower Swan Creek watershed and the Western Basin of Lake Erie. This geographic area encompasses ecologically relevant aquatic habitat features and the relevant portions of associated populations of Federally listed aquatic species or designated critical habitat that could be affected by building and operation of the proposed Fermi 3.

Impacts on aquatic resources can result from changes in habitat availability or quality, degradation of water quality, and increased mortality of organisms. Activities and environmental changes that may contribute to cumulative impacts on aquatic resources within the geographic area of interest include building and operating the proposed Fermi 3, operation of other power plants (including the existing Fermi 2), discharge of treated wastewater, surface water runoff, increased urban development, agricultural activities, commercial and recreational fisheries, introduced invasive species, and global climate change. Human activities have resulted in considerable changes in the Lake Erie aquatic ecosystem during the past century. These changes have resulted from many causes, including overfishing, introduction and expansion of invasive exotic species, nutrient enrichment, dredging, degradation of tributary conditions and other habitat features, and introduction of contaminants.

As described in previous sections, building Fermi 3 and the associated transmission lines would have no effect on Federally listed aquatic species in the Western Basin of Lake Erie or in the lower Swan Creek watershed. If the BMPs identified in previous sections of this BA are implemented, the impacts on the aquatic environment from Fermi 3 building activities, including development of associated transmission lines, would be negligible and discountable and should not appreciably or detectably increase cumulative impacts on Federally listed aquatic species within the geographic area of interest. Therefore, there would be little or no contribution to the cumulative impacts on these species due to the building of Fermi 3. Thus, even where other development projects that occur along the shores of Lake Erie's Western Basin or within watersheds that drain into the Western Basin would contribute to the impacts on Federally listed aquatic species within the geographic area of interest, the contribution of building and operating Fermi 3 to the overall cumulative level of impact would be negligible.

The Lake Erie aquatic ecosystem is also affected by urbanization, industrialization, and agriculture. Development of Fermi 3 and other proposed projects in the region (see Table 7-1 in the draft EIS), could result in increased population and additional urbanization, with subsequent impacts on aquatic resources within the Western Basin of Lake Erie or in the lower Swan Creek watershed. Increased urbanization within the region could affect aquatic resources by increasing the amount of impervious surface, non-point source pollution, and water use, and by altering riparian and in-stream habitat and existing hydrology patterns. Agricultural development within the basin introduces large amounts of sediment to Lake Erie (LaMP Work Group 2008). Overall, the contribution of building and operating Fermi 3 to the cumulative effects from such development-related effects within the Western Basin of Lake Erie or the Lower Swan Creek watershed is expected to be negligible.

There are five operational power plants within the geographic area of interest, including Fermi 2 (located on the Fermi site), the Detroit Edison Monroe Power Plant (6 mi southwest of the Fermi site), the J.R. Whiting Power Plant (14 mi south-southwest of the Fermi site), the Bayshore Power Plant (20 mi south-southwest of the Fermi site), and the Davis-Besse Nuclear Plant (27 mi southeast of the Fermi site). All of these power plants withdraw cooling water from and discharge heated effluent into the Western Basin of Lake Erie. Fermi 2 and Davis-Besse use closed-cycle cooling; the Whiting, Bayshore, and Monroe Power Plants employ once-through cooling. Withdrawing cooling water has a potential to affect aquatic organisms through impingement and entrainment. If the organisms being entrained or impinged at different power plants are members of the same populations, the impacts on those populations would be cumulative. Because the water intakes for Fermi 2 and Fermi 3 would be located in close proximity within the intake bay, it is estimated that the combined operation of the Fermi 2 and Fermi 3 facilities would effectively double the water intake and would likely increase entrainment and impingement rates of aquatic organisms in the immediate vicinity of the intake bay compared to the operation of Fermi 2 alone (Detroit Edison 2011a). However, as described in Section 5.2.1, Fermi 3 is not expected to entrain or impinge the free-living life stages of the listed mussel species identified in Table 1-1, although small numbers of host fish species for these mussels could be entrained or impinged. Overall, entrainment and impingement effects of Fermi 3 on Federally listed aquatic species would be undetectable.

Discharge of heated cooling water from power plants also has the potential to affect survival and growth of organisms by altering ambient water temperatures. In most cases, thermal plumes from power plants discharging into Lake Erie would be expected to affect relatively small areas, and the plumes from Fermi 3 and the other power plants in the Western Basin are not expected to overlap (including the thermal plumes for Fermi 2 and the proposed Fermi 3). As described in Section 5.2.1, thermal effects associated with cooling water discharge during operation of

Fermi 3 would be unlikely to affect mussels, because the discharge ports would direct water upward and not toward the lake bottom. As a consequence, the contribution of Fermi 3 to cumulative effects of thermal discharges on the northern riffleshell, the rayed bean, and the snuffbox mussel within the Western Basin of Lake Erie would be negligible.

Adverse cumulative effects on water quality associated with other projects and activities (e.g., agriculture, storm water runoff, sewage and wastewater treatment facilities) in the Western Basin of Lake Erie and the lower Swan Creek watershed are likely to be significant overall; however, the incremental contribution of Fermi 3 operations to the cumulative impact would be minor.

Dredging occurs in many locations within the Western Basin of Lake Erie and has the potential to affect aquatic biota and habitats through disturbance of benthic habitats, increased turbidity, the suspension and deposition of sediment, introduction of contaminants, and other changes in water quality. The potential for dredging to affect aquatic habitats and biota depends upon the uniqueness and sensitivity of the habitat that would be disturbed by dredging or by disposal of dredged sediments, the types of organisms present in the areas that would be affected, and the size of the area. Although some small areas of a the Fermi site would be affected by dredging in order to build and operate Fermi 3, the dredged materials would be disposed of in onsite disposal areas, not in the open waters of Lake Erie. Whereas cumulative impacts of all dredging activities within the Western Basin of Lake Erie could have small to moderate impacts on aquatic resources in general and, potentially, on some listed aquatic species, there would be no detectable incremental contribution to the overall cumulative impact on the northern riffleshell, rayed bean, or snuffbox mussel due to dredging at the Fermi site because of the minor and infrequent dredging that needs to occur and because the three species of concern are likely not present in the geographic area of interest.

The presence of invasive non-native species is one of the major stressors affecting the Lake Erie ecosystem (LaMP Work Group 2008), including the survival of listed mussel species. These species may prey on native species or compete with them for limited resources, thereby altering the structure of aquatic ecosystems. For example, invasions by quagga (*Dreissena rostriformis bugensis*) and zebra mussels (*Dreissena polymorpha*) have affected ecosystem conditions in Lake Erie by altering nutrient conditions and competing with other species that feed on phytoplankton and zooplankton. Increases in these species have been implicated in the declines of native freshwater mussels. Invasive nuisance organisms that have been found or are presumed to occur in Lake Erie in the vicinity of the Fermi site include the fishhook water flea (*Cercopagis pengoi*), the spiny water flea (*Bythotrephes longimanus*), quagga and zebra mussels, the sea lamprey (*Petromyzon marinus*), and the round goby (*Neogobius melanostomus*). These species are not considered abundant in the vicinity of the Fermi site. Although the cumulative impacts of invasive species on the Lake Erie ecosystem are undoubtedly significant, the building and operation of Fermi 3 would not be expected to measurably promote expansion of populations of invasive species. Thus, the incremental contribution of Fermi 3 to cumulative impacts on Federally listed aquatic species from invasive species would be negligible.

The EPA's Great Lakes National Program Office has initiated the Great Lakes Restoration Initiative to address environmental issues in five topical areas: cleaning up toxics and areas of concern, combating invasive species, promoting nearshore health by protecting watersheds from polluted runoff, restoring wetlands and other habitats, and tracking progress and working with strategic partners. It is expected that this long-term initiative would address some water quality and non-native species concerns that contribute to cumulative impacts of aquatic resources in the area of interest.

The review team is also aware that potential climate changes together with reactor operations could affect water quality and aquatic ecosystems. A study conducted by the U.S. Global Change Research Program (USGCRP 2009) projected that during the operating license period for Fermi 3 (estimated to be 2020 to 2060), changes in the region's climate would include a 3–4°F increase in the average temperature, slightly increased precipitation in the winter and spring, more intense rainstorms throughout the year, and a drop of 1–1.5 ft in the average water levels in Lake Erie. These changes could lead to increased erosion and sediment loading in tributaries and in Lake Erie. It is expected that as temperatures increase and water quality changes due to climate change, a long-term shift could occur in the aquatic species assemblages present within the region (USGCRP 2009). With increases in evaporation rates and longer periods between rainfalls, the likelihood of drought will increase and water levels in rivers, streams, and wetlands are likely to decline (USGCRP 2009), thereby reducing the availability of some aquatic habitats. It is also predicted that reduced summer water levels are likely to reduce the recharge of groundwater, causing small streams to dry up and potentially reducing habitat needed by native aquatic biota such as freshwater mussels. The size of coastal wetland areas that are important for specific life stages of many aquatic organisms within the region could also be affected. Such changes in aquatic species assemblages are likely to be further affected by invasions of non-native species that could thrive under warmer conditions. USGCRP (2009) also predicts that in some lakes increased water temperatures could lead to an earlier and longer period in summer during which mixing of the relatively warm surface lake water with the colder water below is reduced, potentially increasing the risk of developing oxygen-poor zones that could result in increased mortality of fish and other aquatic organisms.

The review team concludes that, with projected climate change and past, present, and reasonably foreseeable future actions in the Lower Swan Creek watershed and the Western Basin of Lake Erie, cumulative impacts on aquatic resources could alter noticeably but not destabilize important attributes of the aquatic resource. However, there would be no detectable incremental contribution to the overall cumulative impact related to global climate change to the northern riffleshell, the rayed bean, or the snuffbox mussel from building and operating Fermi 3. The three species are not known from the site or environs, the habitat is unsuitable for their colonization, and the contribution of Fermi 3 to the furtherance of global climate change would be negligible.

7.0 Conclusions

The potential impacts of building and operating the proposed Fermi 3 project, including the associated offsite transmission lines, on Federally listed terrestrial and aquatic species, including species that are candidates for Federal listing are identified in Table 7-1. The known and probable distributions of these species and the potential ecological impacts of building and operation on the species, their habitats, and the species they interact with have been considered in this BA. Building and operating the subject facilities at the Fermi site would not affect any critical habitat listed under the ESA because no designated critical habitat occurs in the vicinity of the Fermi site or along the route for the proposed transmission lines.

Building and operating the proposed Fermi 3 facilities is not likely to adversely affect terrestrial species listed under the ESA, including candidates for Federal listing, if Detroit Edison meets the conditions stated in Section 5.2. The Indiana bat may be affected but is unlikely to be adversely affected because of the lack of suitable habitat for the species on the Fermi site. The eastern massasauga rattlesnake is unlikely to occur on the Fermi site and would not be affected by building and operating the proposed Fermi 3 facilities. The Karner blue butterfly is unlikely to occur in the project area and therefore would not be affected by building and operating the proposed Fermi 3 facilities. The Mitchell's satyr butterfly is unlikely to occur in the project area and therefore would not be affected by building and operating the proposed Fermi 3 facilities. The American burying beetle is presumed extirpated from Michigan and therefore would not be affected by building and operating the proposed Fermi 3 facilities. Habitat for the eastern prairie fringed orchid is not present on the Fermi site. Therefore, building and operating the proposed Fermi 3 facilities would not affect the eastern prairie fringed orchid.

Clearing forest vegetation for new ROWs for proposed transmission lines (a preconstruction activity that is not a part of the NRC action) and operation of the transmission line would not likely adversely affect individuals of terrestrial species indicated in Table 7-1 if Detroit Edison and ITC*Transmission* meet the conditions stated in Section 5.2.

Habitat along the offsite transmission line corridor has not been surveyed for potential Indiana bat habitat and it is possible suitable habitat currently exists. Because Detroit Edison can avoid adverse impacts, building the proposed transmission lines may affect, but is not likely to adversely affect, the Indiana bat. The eastern massasauga rattlesnake could occur within the transmission line corridor. Because Detroit Edison has the ability to reduce impacts on the snake by simple management efforts, building the transmission line may affect, but is unlikely to adversely affect, the eastern massasauga rattlesnake.

Table 7-1. Summary of Potential Effects on Federally Listed Threatened and Endangered Species and Species That Are Candidates for Federal Listing from Building and Operation of Proposed Fermi 3 and Associated Transmission Lines

Common Name	Scientific Name	Status[a]	Determination
Terrestrial Species			
Mammals			
Indiana bat	*Myotis sodalis*	E	May affect; not likely to adversely affect
Reptiles			
Eastern massasauga rattlesnake	*Sistrurus catenatus catenatus*	C	May affect; not likely to adversely affect
Insects			
Karner blue butterfly	*Lycaeides melissa samuelis*	E	No effect
Mitchell's satyr	*Neonympha mitchellii mitchellii*	E	No effect
American burying beetle	*Nicrophorus americanus*	E	No effect
Vascular Plants			
Eastern prairie fringed orchid	*Platanthera leucophaea*	T	May affect; not likely to adversely affect
Aquatic Species			
Mollusks (Mussels)			
Northern riffleshell	*Epioblasma torulosa rangiana*	E	No effect
Rayed bean	*Villosa fabalis*	E	No effect
Snuffbox mussel	*Epioblasma triquetra*	E	No effect

Source: FWS (2009a)
(a) T = Federal threatened; E = Federal endangered; C = Federal candidate.

The Karner blue butterfly is unlikely to occur in the transmission line corridor. Building the transmission line, therefore, would have no effect on the Karner blue butterfly. The Mitchell's satyr butterfly is unlikely to occur in the project area and, therefore, would not be affected by building and operating the proposed transmission line. The American burying beetle is considered extirpated from Michigan and therefore would not be affected by building and operating the transmission line. The eastern prairie fringed orchid could occur within the transmission line corridor. Because Detroit Edison has the ability to reduce impacts on the eastern prairie fringed orchid by simple management efforts, building the transmission line may affect, but is unlikely to adversely affect, the eastern prairie fringed orchid.

Building and operating the proposed Fermi 3 facilities is also unlikely to affect any Federally listed aquatic species. The northern riffleshell is likely not present in waters of Lake Erie adjacent to the Fermi site. This species is also unlikely to be present in streams that would be crossed by the associated transmission lines. Streams would be spanned without placing towers or other structures in the stream channel and BMPs would be implemented during building and operation of transmission lines to limit the potential for sediment or contaminants to enter waterways. Based on this review, the NRC and the USACE conclude that building and operation of Fermi 3 or the associated transmission lines would not affect the northern riffleshell.

Based on the absence of observations of specimens in available survey data, it is very unlikely that the rayed bean or the snuffbox mussel are present in the vicinity of the Fermi site. In addition, an assessment of habitat conditions indicates that the substrates in the areas that would be disturbed by building of the cooling water intake structure, barge slip and discharge structure for Fermi 3 are not appropriate for these species. Therefore, these species would not be affected by building or operating Fermi 3. Although it is highly unlikely that either of these two species are present in stream drainages crossed by the proposed transmission lines there would be no direct impacts because the streams would be spanned without placing towers or other structures in the stream channel. In addition, BMPs would be implemented during building and operation of transmission lines to limit the potential for sediment or contaminants to enter waterways. Based on this review, the NRC and the Corps conclude that the building and operation of Fermi 3 or the associated transmission lines would not affect the rayed bean or the snuffbox mussel.

8.0 References

10 CFR Part 52. Code of Federal Regulations, Title 10, *Energy*, Part 52, "Licenses, Certifications, and Approvals for Nuclear Power Plants."

10 CFR Part 50. Code of Federal Regulations, Title 10, *Energy*, Part 50, "Domestic Licensing of Production and Utilization Facilities.

10 CFR Part 51. Code of Federal Regulations, Title 10, *Energy*, Part 51, "Environmental Protection Regulations for Domestic Licensing and Related Regulatory Functions.

40 CFR Part 122. Code of Federal Regulations, Title 40, *Protection of Environment*, Part 122, "EPA Administered Permit Programs: the National Pollutant Discharge Elimination System."

58 FR 5638. January 22, 1993. "Endangered and Threatened Wildlife and Plants; Determination of Endangered Status for the Northern Riffleshell Mussel (*Epioblasma torulosa rangiana*) and the Clubshell Mussel (*Pleurobema clava*)." *Federal Register*. U.S. Fish and Wildlife Service.

72 FR 57416. October 9, 2007. "Limited Work Authorizations for Nuclear Power Plants." *Federal Register*. U.S. Nuclear Regulatory Commission.

72 FR 57432. October 9, 2007. "Limited Word Authorizations for Nuclear Power Plants, Final Rule." *Federal Register*. U.S. Nuclear Regulatory Commission.

77 FR 8632. February 14, 2012. "Endangered and Threatened Wildlife and Plants; Determination of Endangered Status for the Rayed Bean and Snuffbox Mussels Throughout Their Ranges." *Federal Register*. U.S. Fish and Wildlife Service.

AECOM. 2009. *Aquatic Ecology Characterization Report: Detroit Edison Company Fermi 3 Project*. November.

Carman, S.M. 2001a. *Special Animal Abstract for* Villosa fabalis *(Rayed Bean)*. Michigan Natural Features Inventory, Lansing, Michigan. Available at http://web4.msue.msu.edu/ mnfi/abstracts/zoology/Villosa_fabalis.pdf. Accessed August 5, 2010.

Carman, S.M., and R.R. Goforth. 2000a. *Special Animal Abstract for* Epioblasma torulosa rangiana *(Northern Riffleshell Mussel)*. Michigan Natural Features Inventory, Lansing, Michigan. Available at http://web4.msue.msu.edu/mnfi/abstracts/zoology/Epioblasma_ torulosa_rangiana.pdf. Accessed August 4, 2010.

Carman, S.M., and R.R. Goforth. 2000b. *Special Animal Abstract for* Epioblasma triquetra *(Snuffbox).* Michigan Natural Features Inventory, Lansing, Michigan. Available at http://web4.msue.msu.edu/mnfi/abstracts/zoology/Epioblasma_triquetra.pdf. Accessed August 9, 2010.

Clean Water Act (CWA). 33 USC 1251, *et seq.* (also referred to as the Federal Water Pollution Control Act).

Detroit Edison Company (Detroit Edison). 1977. *Fermi Atomic Power Plant Unit 2-Applicants Environmental Report, Operating License Stage.* Vol. 1–3, August.

Detroit Edison Company (Detroit Edison). 2009a. Letter from P.W. Smith (Director of Nuclear Development—Licensing, Detroit Edison) to U.S. Nuclear Regulatory Commission dated November 23, 2009, "Subject: Detroit Edison Company Response to NRC Requests for Additional Information Related to the Environmental Review." Accession No. ML093380331.

Detroit Edison Company (Detroit Edison). 2009b. Letter from P.W. Smith (Director of Nuclear Development—Licensing, Detroit Edison) to U.S. Nuclear Regulatory Commission dated July 31, 2009, "Subject: Detroit Edison Company Response to NRC Requests for Additional Information Related to the Environmental Review." Accession No. ML092290662.

Detroit Edison Company (Detroit Edison). 2010. Letter from P.W. Smith (Director of Nuclear Development—Licensing, Detroit Edison) to U.S. Nuclear Regulatory Commission dated March 30, 2010, "Subject: Detroit Edison Company Response to NRC Requests for Additional Information Related to the Environmental Review." Accession No. ML100960472.

Detroit Edison Company (Detroit Edison). 2011a. *Fermi 3 Combined License Application, Part 3: Environmental Report.* Revision 2, Detroit, Michigan. February. Accession No. ML110600498.

Detroit Edison Company (Detroit Edison). 2011b. Letter from P.W. Smith (Director of Nuclear Development—Licensing, Detroit Edison) to U.S. Nuclear Regulatory Commission dated January 10, 2011, "Subject: Updates to the Fermi 3 Combined License Application (COLA) Reflecting Changes to the Fermi Site Layout." Accession No. ML110280343.

Detroit Edison Company (Detroit Edison). 2011c. Memorandum Re: Indiana Bat Reconnaissance Surveys at the Fermi 3 Site. From Ed Shadrick (Black and Veatch) to Randy Westmoreland (Detroit Edison) dated August 11, 2011. Accession No. 11286A089.

Endangered Species Act (ESA) of 1973. 16 USC 1531, *et seq.*

Fisher, B. 2011. Personal communication from B. Fisher (U.S. Fish and Wildlife Service, East Lansing, MI) to D. Weeks (Ecology and Environment, Inc.). March 11.

Francis, J., and J. Boase. 2007. *A Fisheries Survey of Selected Lake Erie Coastal Marshes in Michigan*, 2005, Michigan Department of Natural Resources and U.S. Fish and Wildlife Service.

Gustavson, K., and J. Ohren. 2005. *Stony Creek Watershed Management Plan.* Michigan Department of Natural Resources and Environment. June.

Hoving, C. 2010. Personal communication from C. Hoving (Endangered Species Coordinator, Michigan Department of Natural Resources) to David Weeks (Ecology and Environment, Inc.), July 8.

Lake Erie Lakewide Management Plan (LaMP) Work Group. 2008. *Lake Erie Lakewide Management Plan.* Prepared by the Lake Erie LaMP Management Committee, Environment Canada, and the U.S. Environmental Protection Agency. Available at http://www.epa.gov/greatlakes/lamp/le_2008/index.html. Accessed August 23, 2010.

Lawler, Matusky, and Skelly Engineers. 1993. *Fish Entrainment and Impingement Study, Fermi 2 Power Plant, October 1991–September 1992.* February.

Lee, Y. 2000. *Special Animal Abstract for* Neonympha mitchellii mitchellii *(Mitchell's Satyr Butterfly).* Michigan Natural Features Inventory, Lansing, Michigan.

Lee, Y., and J.T. Legge. 2000. *Special Animal Abstract for* Sistrurus catenatus catenatus *(Eastern Massasauga).* Michigan Natural Features Inventory, Lansing, Michigan.

Michigan Department of Environmental Quality (MDEQ). 1998. *A Biological Survey of Stony Creek and its Tributaries, Amos Palmer Drain, and Ross Drain, Monroe County, July 1997.* MI/DEQ/SWQ-97/087. February.

Michigan Department of Natural Resources (MDNR). 2005. *Wildlife Action Plan.* Available at http://www.michigan.gov/dnrwildlifeactionplan. Accessed February 11, 2010.

Michigan Natural Features Inventory (MNFI). 2007a. *Rare Species Explorer* (Web Application). Available online at http://web4.msue.msu.edu/mnfi/explorer. Accessed June 15, 2011.

Michigan Natural Features Inventory (MNFI). 2007b. "Myotis sodalis *Indiana Bat." Rare Species Explorer.* Available at http://web4.msue.msu.edu/mnfi/explorer/species.cfm?id=11426. Accessed February 12, 2010.

Michigan Natural Features Inventory (MNFI). 2007c. "Eastern Prairie Fringed Orchid." *Rare Species Explorer.* Available at http://web4.msue.msu.edu/mnfi/explorer/species.cfm?id=15534. Accessed June 14, 2010.

National Environmental Policy Act of 1969, as amended (NEPA). 42 USC 4321, *et seq.*

NatureServe. 2009. *NatureServe Explorer: An Online Encyclopedia of Life.* NatureServe, Arlington, Virginia. Available at http://www.natureserve.org/explorer. Accessed June 27, 2010.

NUS Corporation. 1974. *1973–74 Annual Report of the Terrestrial Ecological Studies at the Fermi Site.* Prepared for Detroit Edison Company, March 1974.

Rabe, M.L. 2001. *Special Animal Abstract for* Lycaeides melissa samuelis *(Karner Blue).* Michigan Natural Features Inventory. Lansing, Michigan.

Rivers and Harbors Appropriation Act of 1899. 33 USC 403, *et seq.*

U.S. Fish and Wildlife Service (FWS). 2002. *Status Assessment Report for the Rayed Bean,* Villosa fabalis*, Occurring in the Mississippi River and Great Lakes Systems (U.S. Fish and Wildlife Service Regions 3, 4, and 5, and Canada).* Ohio River Valley Ecosystem Team, Mollusk Subgroup, U.S. Fish and Wildlife Service, Asheville, North Carolina, September.

U.S. Fish and Wildlife Service (FWS). 2007. *Indiana Bat (*Myotis sodalis*) Draft Recovery Plan: First Revision.* Available at http://www.mcrcc.osmre.gov/MCR/Resources/bats/pdf/IN%20BAT%20DRAFT%20PLAN%20apr07.pdf. Accessed November 1, 2011.

U.S. Fish and Wildlife Service (FWS). 2009a. Letter from Craig Czarnecki (Field Supervisor, East Lansing Field Office) to Gregory Hanchett of the Nuclear Regulatory Commission dated January 28, 2009.

U.S. Fish and Wildlife Service (FWS). 2009b. *Range-wide Indiana Bat Protection and Enhancement Plan Guidelines.* Accession No. ML112650059.

U.S. Fish and Wildlife Service (FWS). 2010. *National Wetlands Inventory.* Available at http://www.fws.gov/wetlands/Data/Mapper.html. Accessed July 15, 2010.

U.S. Fish and Wildlife Service (FWS). 2011a. Endangered Species Program website. Available at http://www.fws.gov/endangered/species/us-species.html. Accessed June 15, 2011

U.S. Fish and Wildlife Service (FWS). 2011b. "Species Profile: Eastern Massasauga *(Sitrurus catenatus)."* *Environmental Conservation Online System.* Available at http://ecos.fws.gov/speciesProfile/profile/speciesProfile.action?spcode=C03P. Accessed June 15, 2011.

U.S. Fish and Wildlife Service (FWS). 2011c. "Species Profile: Mitchell's Satyr Butterfly *(Neonympha mitchellii mitchellii)*." *Environmental Conservation Online System.* Available at http://ecos.fws.gov/speciesProfile/profile/speciesProfile.action?spcode=I00K. Accessed June 15, 2011.

U.S. Fish and Wildlife Service (FWS). 2011d. *Indiana Bat (*Myotis sodalis*) Bat Survey Protocol for Assessing Use of Potential Hibernacula.* Available at http://www.fws.gov/Midwest/////// Endangered/mammals/inba/inba_srvyprtcl.html.

U.S. Global Change Research Program (USGCRP). 2009. *Global Climate Change Impacts in the United States.* T.R. Karl, J.M. Melillo, and T.C. Peterson (eds.). Cambridge University Press, New York. Available at http://downloads.globalchange.gov/usimpacts/pdfs/climate-impacts-report.pdf. Accessed August 9, 2010.

U.S. Nuclear Regulatory Commission (NRC). 1996. *Generic Environmental Statement for License Renewal of Nuclear Power Plants.* NUREG-1437. Washington, D.C.

U.S. Nuclear Regulatory Commission (NRC). 2000. *Environmental Standard Review Plan – Review Plans for Environmental Reviews for Nuclear Power Plants.* NUREG-1555, Washington, D.C. Includes 2007 updates.

Appendix G

Supporting Documentation on the Radiological Dose Assessment for Fermi 3

Appendix G

Supporting Documentation on the Radiological Dose Assessment for Fermi 3

The U.S. Nuclear Regulatory Commission (NRC) staff performed an independent dose assessment of the radiological impacts resulting from normal operation of the new nuclear unit at the Detroit Edison Enrico Fermi Atomic Power Plant (Fermi) site. The results of this assessment are presented in this appendix and are compared to the results from the Detroit Edison Company (Detroit Edison) found in this environmental impact statement (EIS) in Section 5.9, Radiological Impacts of Normal Operations. This appendix is divided into four sections: (1) dose estimates to the public from liquid effluents, (2) dose estimates to the public from gaseous effluents, (3) cumulative dose estimates, and (4) dose estimates to biota from liquid and gaseous effluents.

G.1 Dose Estimates to the Public from Liquid Effluents

The NRC staff used the dose assessment approach specified in Regulatory Guide 1.109 (NRC 1977) and the LADTAP II computer code (Strenge et al. 1986) to estimate doses to the maximally exposed individual (MEI) and population from the liquid effluent pathway of the proposed Enrico Fermi Unit 3 (Fermi 3).

G.1.1   Scope

Doses from the proposed Fermi 3 to the MEI were calculated and compared to regulatory criteria for the following:

- *Total Body*. Dose was the total for all pathways (i.e., drinking water, fish and shellfish consumption, shoreline usage, swimming exposure, boating) with the highest value for the adult, teen, child, or infant compared to the 3 millirem (mrem)/year (yr) per reactor dose design objective in Title 10 of the Code of Federal Regulations (CFR), Part 50, Appendix I.

- *Organ*. Dose was the total for each organ for all pathways (i.e., drinking water, fish and shellfish consumption, shoreline usage, swimming exposure, boating) with the highest value for the adult, teen, child, or infant compared to the 10 mrem/yr per reactor dose design objective specified in 10 CFR Part 50, Appendix I.

The NRC staff reviewed the assumed exposure pathways and the input parameters and values used by Detroit Edison (2011) for appropriateness, including references made to the General

Electric-Hitachi Nuclear Energy Americas, LLC (GEH) Economic Simplified Boiling Water Reactor (ESBWR) Design Control Document (GEH 2010). Default values from Regulatory Guide 1.109 (NRC 1977) were used when site-specific input parameters were not available from Detroit Edison. The staff concluded that the assumed exposure pathways were reasonable and that the input parameters and values used by Detroit Edison were appropriate.

G.1.2   Resources Used

To calculate doses to the public from liquid effluents, the NRC staff used a personal computer (PC) version of the LADTAP II code entitled NRCDOSE, Version 2.3.10 (Chesapeake Nuclear Services, Inc. 2008) obtained through the Oak Ridge Radiation Safety Information Computational Center (RSICC).

G.1.3   Input Parameters

Table G-1 provides a listing of the major parameters used in calculating dose to the public from liquid effluent releases during normal operation.

G.1.4   Comparison of Results

The NRC staff compared the results documented in the Environmental Report (ER) (i.e., Detroit Edison 2011) with the results calculated by the NRC. Doses calculated for the MEI and population by the NRC staff confirmed the doses calculated by Detroit Edison.

For calculating the population dose from liquid effluents, the population distribution used by Detroit Edison was for year 2060, 10 years beyond the anticipated operating license. However, Environmental Standard Review Plan (ESRP) Section 5.4.1 (NRC 2000) instructs the NRC staff to use the "projected population for 5 years from the time of the licensing action under consideration." Assuming the combined license (COL) licensing action occurred in year 2010 and adding 5 years yields year 2015. Using the population projections from ER Tables 2.5-10 and 2.5-12 (Detroit Edison 2011) (summarized in Table G-2) yields a population estimate for the year 2015 of 5,971,392. This population estimate is significantly smaller than the 2060 projected population (7,713,709), so the doses calculated by Detroit Edison are conservatively high. For comparability, NRC staff also used the 2060 population estimate. Doses for the year 2015 would be lower by a factor of 1.29 than those reported below.

G.2 Dose Estimates to the Public from Gaseous Effluents

The NRC staff used the dose assessment approach specified in Regulatory Guide 1.109 (NRC 1977) and the XOQDOQ and GASPAR II computer codes (Sagendorf et al. 1982;

Table G-1. Parameters Used in Calculating Dose to the Public from Liquid Effluent Releases

Parameter	Staff Value		Comments
New unit liquid effluent source term (curie [Ci]/yr)[a][b]	H-3	1.40×10^1	Values from GEH ESBWR Design Control Document (DCD) Table 12.2-19b for a single unit (GEH 2010).
	Na-24	4.19×10^{-3}	
	P-32	3.51×10^{-4}	
	Cr-51	1.10×10^{-2}	
	Mn-54	1.30×10^{-4}	
	Mn-56	1.00×10^{-3}	
	Fe-55	1.90×10^{-3}	
	Fe-59	6.00×10^{-5}	
	Co-58	3.70×10^{-4}	
	Co-60	7.51×10^{-4}	
	Cu-64	1.00×10^{-2}	
	Zn-65	3.70×10^{-4}	
	Zn-69m	7.51×10^{-4}	
	Br-83	1.00×10^{-4}	
	Sr-89	1.90×10^{-4}	
	Sr-90	1.00×10^{-5}	
	Sr-91	9.51×10^{-4}	
	Y-91	1.20×10^{-4}	
	Sr-92	2.30×10^{-4}	
	Y-92	8.70×10^{-4}	
	Y-93	1.00×10^{-3}	
	Zr-95	1.00×10^{-5}	
	Nb-95	1.00×10^{-5}	
	Mo-99	2.50×10^{-3}	
	Tc-99m	4.60×10^{-3}	
	Ru-103	4.00×10^{-5}	
	Ru-105	1.30×10^{-4}	
	Te-129m	7.00×10^{-5}	
	Te-131m	8.00×10^{-5}	
	Te-132	1.00×10^{-5}	
	I-131	6.19×10^{-3}	
	I-132	9.30×10^{-4}	
	I-133	3.00×10^{-2}	
	I-134	4.00×10^{-5}	
	I-135	7.11×10^{-3}	
	Cs-134	5.70×10^{-4}	
	Cs-136	3.51×10^{-4}	
	Cs-137	1.50×10^{-3}	
	Ba-139	3.00×10^{-5}	

Table G-1. (contd)

Parameter	Staff Value	Comments
	Ba-140 6.89×10^{-4}	
	Ce-141 6.00×10^{-5}	
	La-142 2.00×10^{-5}	
	Ce-143 3.00×10^{-5}	
	Pr-143 7.00×10^{-5}	
	W-187 2.00×10^{-4}	
	Np-239 9.30×10^{-3}	
Discharge flow rate (cubic feet [ft^3]/second [sec])	0.234	Site-specific value from Table 5.4-1 of the ER (Detroit Edison 2011).
Source term multiplier	1	Single-unit source term.
Site type	Fresh water	Discharge is to the freshwater Lake Erie.
Reconcentration model	No impoundment	Site-specific value from Table 5.4-1 of the ER (Detroit Edison 2011).
Impoundment total volume (ft^3)	0	Set to zero for "no impoundment" model (Strenge et al. 1986).
Shore width factor	0.3	Suggested value for lake (NRC 1977; Strenge et al. 1986).
Dilution factor at discharge location	115	Blowdown flow rate divided by discharge flow rate from Table 5.4-1 of the ER (Detroit Edison 2011).
Dilution factors after discharge		Site-specific value from Table 5.4-1 of the ER (Detroit Edison 2011).
Aquatic food and boating	100	
Shoreline and swimming	45	
Drinking water	67 (MEI), 100 (population)	
Transit time (hour [hr])		Site-specific value from Table 5.4-1 of the ER (Detroit Edison 2011).
Drinking water	22.6 (MEI), 24 (population)	
Boating and swimming	10.6	
Fish and invertebrate	24	

Consumption and usage factors for adults, teens, children, and infants	Shoreline recreational usage (hr/yr)
	  12 (adult)
	  67 (teen)
	  14 (child)
	  0 (infant)
		Site-specific values from Table 5.4-2 of the ER (Detroit Edison 2011) and LADTAP II code default values (NRC 1977; Strenge et al. 1986).

Table G-1. (contd)

Parameter	Staff Value	Comments
Drinking water usage (liters [L]/yr) 730 (adult) 510 (teen) 510 (child) 330 (infant) Fish consumption (kilograms [kg]/yr) 21 (adult) 16 (teen) 6.9 (child) 0 (infant)		
Total 50-mile (mi) population	7,713,709	Site-specific value from Table 5.4-1 of the ER (Detroit Edison 2011).
Total 50-mi sport fishing harvest (kg/yr)	11,450,000	Site-specific value from Table 5.4-1 of the ER (Detroit Edison 2011).
Total 50-mi commercial fishing harvest (kg/yr)	2,070,000	Site-specific value from Table 5.4-1 of the ER (Detroit Edison 2011).
Total 50-mi commercial invertebrate harvest (kg/yr)	33,000,000	Site-specific value from Table 5.4-1 of the ER (Detroit Edison 2011).
Total 50-mi shoreline usage (person-hr/yr)	5,747,850	Calculated using site-specific individual value from Table 5.4-1 and usage factors for average individual from Table 5.4-2 of the ER (Detroit Edison 2011), as well as age distribution from LADTAP II code defaults (NRC 1977; Strenge et al. 1986).
Total 50-mi swimming usage (person-hr/yr)	5,747,850	Calculated using site-specific individual value from Table 5.4-1 and usage factors for average individual from Table 5.4-2 of the ER (Detroit Edison 2011), as well as age distribution from LADTAP II code defaults (NRC 1977; Strenge et al. 1986).
Total 50-mi boating usage (person-hr/yr)	5,747,850	Calculated using site-specific individual value from Table 5.4-1 and usage factors for average individual from Table 5.4-2 of the ER (Detroit Edison 2011), as well as age distribution from LADTAP II code defaults (NRC 1977; Strenge et al. 1986).

(a) To convert Ci/yr to Becquerel (Bq)/yr, multiply the value by 3.7×10^{10}.
(b) Only radionuclides included in Regulatory Guide 1.109 are considered (NRC 1977).

Table G-2. Population Projections from 2000 to 2060 within 50 mi of the Fermi Site

Year	Population Projections[a] within Radii/Distances (mi)							Annual Average Percent Change for the 10-Year Period
	0 to 1 mi[a]	1 to 10 mi	10 to 20 mi	20 to 30 mi	30 to 40 mi	40 to 50 mi	0 to 50 mi[d]	
2000[b]	570	106,166	347,077	1,769,937	2,010,398	1,344,775	5,578,923	Not applicable
2008[c]	1163	112,665	348,369	1,791,988	2,081,615	1,449,117	5,784,917	Not applicable
2020[c]	1153	123,378	351,302	1,831,686	2,198,894	1,624,796	6,131,209	Not applicable
2030[c]	1144	133,239	354,711	1,871,367	2,307,607	1,791,234	6,459,302	0.52
2040[c]	1133	144,031	359,060	1,917,634	2,427,914	1,978,702	6,828,474	0.55
2050[c]	1122	155,853	364,415	1,971,113	2,561,627	2,190,275	7,244,405	0.59
2060[c, d]	1109	168,799	370,858	2,032,503	2,710,898	2,429,542	7,713,709	0.63

Source: Detroit Edison 2011

(a) Population estimates and projections include transient and residential population in the 0- to 10-mi range.
(b) Residential population in 2000, U.S. Census Bureau, decennial census.
(c) The populations for years 2008 through 2060 have been projected by calculating a growth rate using State population projections (by county) as the base.
(d) Population used in GASPAR II population runs (Detroit Edison 2011).

Strenge et al. 1987) to estimate doses to the MEI and to the population within a 50-mi radius of the proposed Fermi 3 site from the gaseous effluent pathway for the proposed unit.

G.2.1   Scope

The NRC staff performed multiple calculations to confirm that Detroit Edison properly accounted for dispersion and deposition from three stack releases to identify the most limiting MEI. The maximum gamma air dose, beta air dose, total body dose, and skin dose from noble gases was calculated at the exclusion area boundary location 0.48 mi north-northwest (NNW) of the proposed Fermi 3 site. The maximum dose from ground exposure was calculated at the exclusion area boundary location 0.48 mi west-northwest (WNW) of the proposed Fermi 3 site. The maximum dose to residents and the MEI from consumption of vegetables was calculated at 0.59 mi NW of the site. The maximum dose from the milk ingestion pathway was calculated at 2.1 mi WNW of the site. The maximum dose from the meat ingestion pathway was calculated at 3.0 mi NNW of the site. The dose to the MEI is estimated as the sum of the maximum doses from each of the following exposure pathways: plume immersion, direct shine from deposited radionuclides, inhalation, ingestion of local farm or garden vegetables, ingestion of locally produced beef, and ingestion of locally produced milk (Detroit Edison 2011).

The NRC staff reviewed the input parameters and values used by Detroit Edison (2011) for appropriateness, including references made to the GEH ESBWR design control document (GEH 2010). Default values from Regulatory Guide 1.109 (NRC 1977) were used when site-specific input parameters were not available. The NRC staff concluded that the assumed exposure pathways and input parameters and values used by Detroit Edison were appropriate. These pathways and parameters were used by the NRC staff in its independent calculations using GASPAR II.

Joint frequency distribution data of wind speed and wind direction by atmospheric stability class for the proposed Fermi 3 site (Detroit Edison 2011) were used as input to the XOQDOQ code (Sagendorf et al. 1982) to calculate long-term average atmospheric dispersion χ/Q and deposition factor D/Q values for routine releases. The NRC staff's independent calculations of χ/Q and D/Q values confirmed the values reported by Detroit Edison in ER Tables 2.7-87 through 2.7-152 (Detroit Edison 2011).

Population doses were calculated for all types of releases (i.e., noble gases, iodines and particulates, and H-3 and C-14) using the GASPAR II code for the following exposure pathways: plume immersion, direct shine from deposited radionuclides, ingestion of vegetables, and ingestion of milk and meat.

G.2.2   Resources Used

To calculate doses to the public from gaseous effluents, the NRC staff used a PC version of the XOQDOQ and GASPAR II codes entitled NRCDOSE Version 2.3.10 (Chesapeake Nuclear Services, Inc. 2008) obtained through the Oak Ridge RSICC.

G.2.3   Input Parameters

Table G-3 provides a listing of the major parameters used in calculating dose to the public from gaseous effluent releases during normal operation. For dose estimation, the gaseous effluent source terms from reactor building, turbine building, and radwaste building were evaluated separately.

G.2.4   Comparison of Doses to the Public from Gaseous Effluent Releases

The NRC staff compared the results documented in the ER (Detroit Edison 2011) for doses from noble gases at the exclusion area boundary with the results calculated by the NRC staff. The doses calculated by the NRC staff confirmed the doses calculated by Detroit Edison.

The NRC staff also compared its estimates of the doses to the MEI to the doses calculated by Detroit Edison. Doses to the MEI were calculated at the nearest residence, nearest garden, nearest milk cow, and nearest beef cattle. The term "nearest" means the location where the individual would receive the highest calculated dose for the specific pathway. The doses calculated by the NRC staff confirmed the doses calculated by Detroit Edison.

G.2.5   Comparison of Results – Population Doses

The NRC staff compared its calculations with the Detroit Edison population dose estimates documented in the ER (Detroit Edison 2011, Table 5.4-7). The NRC staff's calculations for population dose confirmed the Detroit Edison estimates (Detroit Edison 2011, Table 5.4-7) for the new Fermi 3. Both Detroit Edison and NRC staff used population estimates for the year 2060, which is a factor of 1.29 times higher than the population estimated for the year 2015 (and 5 years past the expected licensing action).

G.3 Cumulative Dose Estimates

The NRC staff compared the results documented in the ER (Detroit Edison 2011) for cumulative dose estimates to the MEI with those calculated by the NRC staff. Cumulative dose estimates include doses from all pathways (i.e., direct exposure, liquid effluents, and gaseous effluents) for both the proposed Fermi 3 and the existing Fermi 2 at the Fermi site. These cumulative dose estimates were calculated for comparison to the dose standards of

Table G-3. Parameters Used in Calculating Dose to the Public from Gaseous Effluent Releases

Parameter	NRC Staff Value		Comments
New unit gaseous effluent source term – reactor building (Ci/yr)[a]	Kr-83m	2.30×10^{-3}	Values from GEH ESBWR DCD Table 12.2-16 for a single unit (GEH 2010) and Final Safety Analysis Report (FSAR) Table 12.2-206 (Detroit Edison 2012).
	Kr-85m	2.44×10^{0}	
	Kr-85	2.03×10^{-3}	
	Kr-87	1.22×10^{0}	
	Kr-88	2.45×10^{0}	
	Kr-89	1.22×10^{0}	
	Xe-131m	1.11×10^{-3}	
	Xe-133m	5.14×10^{-3}	
	Xe-133	6.79×10^{1}	
	Xe-135m	3.78×10^{1}	
	Xe-135	7.84×10^{1}	
	Xe-137	1.11×10^{2}	
	Xe-138	4.87×10^{0}	
	I-131	3.46×10^{-2}	
	I-132	2.31×10^{-1}	
	I-133	1.76×10^{-1}	
	I-134	4.06×10^{-1}	
	I-135	2.36×10^{-1}	
	H-3	3.95×10^{1}	
	Na-24	1.59×10^{-4}	
	P-32	4.05×10^{-5}	
	Cr-51	5.00×10^{-3}	
	Mn-54	1.94×10^{-3}	
	Fe-55	1.38×10^{-3}	
	Mn-56	3.24×10^{-4}	
	Co-58	5.35×10^{-4}	
	Co-60	7.03×10^{-3}	
	Fe-59	5.78×10^{-4}	
	Ni-63	1.41×10^{-6}	
	Cu-64	2.03×10^{-4}	
	Zn-65	8.14×10^{-3}	
	Rb-89	5.41×10^{-6}	
	Sr-89	1.95×10^{-4}	
	Sr-90	2.32×10^{-5}	
	Sr-91	2.03×10^{-4}	
	Sr-92	1.32×10^{-4}	
	Y-90	2.41×10^{-6}	
	Y-91	5.14×10^{-5}	
	Y-92	1.03×10^{-4}	

Table G-3. (contd)

Parameter	NRC Staff Value		Comments
	Y-93	2.19×10^{-4}	
	Zr-95	1.36×10^{-3}	
	Nb-95	1.35×10^{-2}	
	Mo-99	8.99×10^{-2}	
	Tc-99m	6.49×10^{-5}	
	Ru-103	5.70×10^{-3}	
	Ru-106	4.32×10^{-6}	
	Rh-103m	1.03×10^{-7}	
	Rh-106	1.41×10^{-10}	
	Ag-110m	4.62×10^{-6}	
	Sb-124	6.76×10^{-5}	
	Te-129m	4.86×10^{-5}	
	Te-131m	1.62×10^{-5}	
	Te-132	4.05×10^{-6}	
	Cs-134	6.52×10^{-3}	
	Cs-136	6.93×10^{-4}	
	Cs-137	8.21×10^{-3}	
	Cs-138	2.30×10^{-5}	
	Ba-140	3.01×10^{-2}	
	La-140	3.78×10^{-4}	
	Ce-141	1.25×10^{-3}	
	Ce-144	4.32×10^{-6}	
	Pr-144	4.86×10^{-9}	
	W-187	3.78×10^{-5}	
	Np-239	2.43×10^{-3}	
New unit gaseous effluent source term – turbine building (Ci/yr)[a]	Kr-83m	3.78×10^{-9}	
	Kr-85m	1.53×10^{1}	
	Kr-85	1.41×10^{2}	
	Kr-87	3.78×10^{1}	
	Kr-88	5.41×10^{1}	
	Kr-89	3.51×10^{2}	
	Xe-131m	4.05×10^{0}	
	Xe-133m	2.19×10^{-5}	
	Xe-133	8.99×10^{2}	
	Xe-135m	2.43×10^{2}	
	Xe-135	4.97×10^{2}	
	Xe-137	6.22×10^{2}	
	Xe-138	6.22×10^{2}	
	I-131	1.90×10^{-1}	

Table G-3. (contd)

Parameter	NRC Staff Value		Comments
	I-132	1.24×10^0	
	I-133	9.21×10^{-1}	
	I-134	2.27×10^0	
	I-135	1.27×10^0	
	H-3	3.24×10^1	
	C-14	1.43×10^1	
	Ar-41	3.78×10^{-2}	
	Cr-51	1.22×10^{-3}	
	Mn-54	8.11×10^{-4}	
	Co-58	1.35×10^{-3}	
	Co-60	1.35×10^{-3}	
	Fe-59	1.35×10^{-4}	
	Zn-65	8.11×10^{-3}	
	Sr-89	8.11×10^{-3}	
	Sr-90	2.70×10^{-5}	
	Zr-95	5.41×10^{-5}	
	Nb-95	8.11×10^{-6}	
	Mo-99	2.70×10^{-3}	
	Ru-103	6.76×10^{-5}	
	Sb-124	1.35×10^{-4}	
	Cs-134	2.70×10^{-4}	
	Cs-136	1.35×10^{-4}	
	Cs-137	1.35×10^{-3}	
	Ba-140	1.35×10^{-2}	
	Ce-141	1.35×10^{-2}	
New unit gaseous effluent source term – radwaste building (Ci/yr)[a]	Kr-89	1.76×10^1	Values from GEH ESBWR DCD Table 12.2-16 for a single unit (GEH 2010) and FSAR Table 12.2-206 (Detroit Edison 2012).
	Xe-133	1.35×10^2	
	Xe-135m	3.24×10^2	
	Xe-135	1.70×10^2	
	Xe-137	5.14×10^1	
	Xe-138	1.22×10^0	
	I-131	9.19×10^{-3}	
	I-132	8.11×10^{-2}	
	I-133	5.95×10^{-2}	
	I-134	1.49×10^{-1}	
	I-135	8.38×10^{-2}	
	Cr-51	9.46×10^{-4}	
	Mn-54	5.41×10^{-3}	
	Co-58	2.70×10^{-4}	

Table G-3. (contd)

Parameter	NRC Staff Value		Comments
	Co-60	9.46×10^{-3}	
	Fe-59	4.05×10^{-4}	
	Zn-65	4.05×10^{-4}	
	Zr-95	1.08×10^{-3}	
	Nb-95	5.41×10^{-6}	
	Mo-99	4.05×10^{-6}	
	Ru-103	1.35×10^{-6}	
	Sb-124	9.46×10^{-5}	
	Cs-134	3.24×10^{-3}	
	Cs-137	5.41×10^{-3}	
	Ba-140	5.41×10^{-6}	
	Ce-141	9.46×10^{-6}	
Population distribution	Tables 2.5-10 and 2.5-12 of the ER (Detroit Edison 2011)		Population distribution used by Detroit Edison and the NRC staff was for year 2060. Note that ESRP Section 5.4.1 requires use of "projected population for 5 years from the time of the licensing action under consideration." Assuming the ESRP licensing action occurred in year 2010, adding 5 years yields year 2015. See discussion of population dose in Section G.2.5.
Wind speed and direction distribution	Table 2.7-63 of the ER (Detroit Edison 2011)		Site-specific data provided by Detroit Edison for time periods from 2003 to 2007.
Atmospheric dispersion factors (sec/cubic meter [m^3])	Tables 2.7-87 through 2.7-95 and Tables 2.7-108 through 2.7-140 of the ER (Detroit Edison 2011)		Site-specific data provided by Detroit Edison for time periods from both 1985 to 1989 and 2003 to 2007.
Ground deposition factors (m^{-2})	Tables 2.7-87 through 2.7-95 and Tables 2.7-108 through 2.7-140 of the ER (Detroit Edison 2011)		Site-specific data provided by Detroit Edison for time periods from both 1985 to 1989 and 2003 to 2007.
Milk production rate within a 50-mi radius of the Fermi site (kg/yr)	6.043×10^8		Site-specific data from Table 5.4-3 provided by Detroit Edison (2011).
Vegetable/fruit production rate within a 50-mi radius of the Fermi site (kg/yr)	9.689×10^9		Site-specific data from Table 5.4-3 provided by Detroit Edison (2011).

Table G-3. (contd)

Parameter	NRC Staff Value	Comments
Meat production rate within a 50-mi radius of the Fermi site (kg/yr)	1.919×10^7	Site-specific data from Table 5.4-3 provided by Detroit Edison (2011).
Pathway receptor locations (direction and distance) – nearest site boundary, vegetable garden, residence, meat animal, milk animal	Tables 2.7-80 through 2.7-86 of the ER (Detroit Edison 2011)	Site-specific data provided by Detroit Edison (2011).
Consumption factors for milk, meat, leafy vegetables, and vegetables	Milk (L/yr) 310 (adult) 400 (teen) 330 (child) 330 (infant) Meat (kg/yr) 110 (adult) 65 (teen) 41 (child) 0 (infant) Leafy vegetables (kg/yr) 64 (adult) 42 (teen) 26 (child) 0 (infant) Vegetables (kg/yr) 520 (adult) 630 (teen) 520 (child) 0 (infant)	Table 5.4-2 of the ER (Detroit Edison 2011) and Regulatory Guide 1.109 (NRC 1977).
Fraction of year that leafy vegetables are grown	0.33	Site-specific value from Table 5.4-3 of the ER (Detroit Edison 2011).
Fraction of year that milk cows are on pasture	0.58	Site-specific value from Table 5.4-3 of the ER (Detroit Edison 2011).
Fraction of year that goats are on pasture	0.67	Site-specific value from Table 5.4-3 of the ER (Detroit Edison 2011)
Fraction of MEI vegetable intake from own garden	0.76	Default value of GASPAR II code (Strenge et al. 1987).
Fraction of milk-cow intake that is from pasture while on pasture	1	Default value of GASPAR II code (Strenge et al. 1987).
Fraction of goat intake that is from pasture while on pasture	1	Default value of GASPAR II code (Strenge et al. 1987).

Table G-3. (contd)

Parameter	NRC Staff Value	Comments
Average absolute humidity over the growing season (g/m^3)	11	Site-specific value from the Detroit Edison (2011), Table 5.4-3.
Average temperature over the growing season (°F)	None	Default value of GASPAR II code (Strenge et al. 1987).
Fraction of year that beef cattle are on pasture	0.58	Site-specific value from Table 5.4-3 of the ER (Detroit Edison 2011).
Fraction of year of beef cattle intake that is from pasture while on pasture	1	Default value of GASPAR II code (Strenge et al. 1987).

(a) To convert Ci/yr to Bq/yr, multiply the value by 3.7×10^{10}.

40 CFR Part 190. The NRC staff's calculations for cumulative dose confirmed the Detroit Edison estimates (Detroit Edison 2011, Table 5.4-8).

G.4 Dose Estimates to the Biota from Liquid and Gaseous Effluents

To estimate doses to the biota from the liquid and gaseous effluent pathways, the NRC staff used the LADTAP II code (Strenge et al. 1986), the GASPAR II code (Strenge et al. 1987), and input parameters supplied by Detroit Edison in its ER (Detroit Edison 2011).

G.4.1   Scope

The NRC staff estimated the doses to biota other than human beings using surrogate species; using the characteristics of surrogate species to represent a range of species is an accepted methodology. Fish, algae, and invertebrate species are used as surrogate aquatic biota species. Muskrats, raccoons, herons, and ducks are used as surrogate terrestrial biota species. The staff recognizes the LADTAP II computer program as an appropriate method for calculating doses to the aquatic biota and for calculating the liquid-pathway contribution to terrestrial biota. The LADTAP II code calculates an internal dose component and an external dose component and sums them for a total body dose. The NRC staff reviewed the input parameters used by Detroit Edison for appropriateness. Default values from Regulatory Guide 1.109 (NRC 1977) were used when site-specific input parameters were not available. The NRC staff concluded that all of the LADTAP II input parameters used by Detroit Edison were appropriate. These parameters were used by the NRC staff in its independent calculations using LADTAP II.

The LADTAP II code calculates only biota doses from the liquid effluent pathway. Terrestrial biota could also be exposed via the gaseous effluent pathway. The gaseous pathway doses would be the same as doses for the MEI calculated using the GASPAR II code. Detroit Edison

(2011) used the MEI doses at 0.25 mi from the release point to estimate onsite biota exposures. To account for the greater proximity of the main body mass of animals to the ground as compared to that of humans, the biota calculation assumed a ground deposition factor twice that used in the human MEI calculation. The gaseous pathway doses are summed and combined with the liquid pathway doses for the representative biota species. The NRC staff used the same approach in its calculations with one exception. The NRC staff included doses from ingestion of vegetation in the gaseous pathway estimates.

G.4.2   Resources Used

To calculate doses to the biota, the NRC staff used a PC version of the LADTAP II and GASPAR II computer codes entitled NRCDOSE Version 2.3.10 (Chesapeake Nuclear Services, Inc. 2008). NRCDOSE was obtained through the Oak Ridge RSICC.

G.4.3   Input Parameters

The NRC staff used the input parameters for LADTAP II and GASPAR II specified in Sections G.2.3 and G.2.4 to calculate biota doses.

G.4.4   Comparison of Results

Table G-4 compares Detroit Edison's biota dose estimates from liquid and gaseous effluents presented in the ER (Detroit Edison 2011, Table 5.4-9) with the NRC staff's estimates. The NRC staff's dose estimates were slightly higher than Detroit Edison's estimates for gaseous pathways because of the addition of the vegetation ingestion pathway.

Table G-4. Comparison of Dose Estimates to Biota from Liquid and Gaseous Effluents for Fermi 3

Biota	Pathway	Detroit Edison (2011, Table 5.4-9) (milliradian [mrad]/yr)	NRC Staff Calculation (mrad/yr)	Percent Difference
Fish	Liquid	2.31	2.31	0
	Gaseous[a]	NA	NA	–
Muskrat	Liquid	14.8	14.8	0
	Gaseous	11.15	12.7	12
Raccoon	Liquid	0.43	0.43	0
	Gaseous	11.15	12.7	12
Heron	Liquid	6.87	6.87	0
	Gaseous	11.15	12.7	12
Duck	Liquid	14.8	14.8	0
	Gaseous	11.15	12.7	12
Algae	Liquid	11.9	11.9	0
	Gaseous[a]	NA	NA	–
Invertebrate	Liquid	7.65	7.65	0
	Gaseous[a]	NA	NA	–

(a) Fish, invertebrate species, and algae would not be exposed to gaseous effluents.

G.5 References

10 CFR Part 50. Code of Federal Regulations, Title 10, *Energy*, Part 50, "Domestic Licensing of Production and Utilization Facilities."

40 CFR Part 190. Code of Federal Regulations, Title 40, *Protection of Environment*, Part 190 "Environmental Radiation Protection Standards for Nuclear Power Operations."

Chesapeake Nuclear Services, Inc. 2008. *NRCDOSE for Windows*. Radiation Safety Information Computational Center, Oak Ridge, Tennessee.

Detroit Edison Company (Detroit Edison). 2011. *Fermi 3 Combined License Application, Part 3: Environmental Report*. Revision 2, Detroit, Michigan. February. Accession No. ML110600498.

Detroit Edison Company (Detroit Edison). 2012. *Fermi 3 Combined License Application, Part 2: Final Safety Analysis Report*. Revision 4, Detroit, Michigan. February. Accession No. ML12095A119.

General Electric-Hitachi Nuclear Energy Americas, LLC (GEH). 2010. *ESBWR Design Control Document – Tier 2, Chapter 12 Radiation Protection.* Revision 9. December. Accession No. ML103440247.

Sagendorf, J.F., J.T. Goll, and W.F. Sandusky. 1982. *XOQDOQ: Computer Program for the Meteorological Evaluation of Routine Effluent Releases at Nuclear Power Stations.* NUREG/CR-2919, Pacific Northwest National Laboratory, Richland, Washington.

Strenge, D.L., R.A. Peloquin, and G. Whelan. 1986. *LADTAP II – Technical Reference and User Guide.* NUREG/CR-4013, Pacific Northwest Laboratory, Richland, Washington.

Strenge, D.L., T.J. Bander, and J.K. Soldat. 1987. *GASPAR II – Technical Reference and User Guide.* NUREG/CR-4653, Pacific Northwest Laboratory, Richland, Washington.

U.S. Nuclear Regulatory Commission (NRC). 1977. *Calculation of Annual Doses to Man from Routine Releases of Reactor Effluents for the Purpose of Evaluating Compliance with 10 CFR Part 50, Appendix I.* Regulatory Guide 1.109, Office of Nuclear Reactor Regulation, Washington, D.C.

U.S. Nuclear Regulatory Commission (NRC). 2000. *Standard Review Plans for Environmental Reviews for Nuclear Power Plants: Environmental Standard Review Plan.* NUREG-1555, Office of Nuclear Reactor Regulation, Washington, D.C. Available at http://www.nrc.gov/reading-rm/doc-collections/nuregs/staff/sr1555. Accessed July 13, 2008.

Appendix H

Authorizations, Permits, and Certifications

Appendix H

Authorizations, Permits, and Certifications

This appendix contains a list (Table H-1) of the environment-related authorizations, permits, and certifications potentially required by Federal, State, regional, local, and affected Native American Tribal agencies related to the combined license for the proposed Enrico Fermi Unit 3 (Fermi 3). The table is adapted from Table 1.2-1 of the Environmental Report (ER) submitted to the U.S. Nuclear Regulatory Commission (NRC) by the applicant, Detroit Edison Company (Detroit Edison).

Table H-1. Authorizations/Permits Required for Combined License

Agency[a]	Authority	Requirement	Activity Covered	Status[b]
Federal Authorizations				
NRC	10 Code of Federal Regulations (CFR) Part 52, Subpart C	Combined License	Construction activities associated with a nuclear power facility.	Submitted September 18, 2008
NRC	10 CFR Part 40	Source Material License	Approval to possess source material.	To be issued as part of COL
NRC	10 CFR Part 70	Special Nuclear Materials License	Approval to possess special nuclear material.	To be issued as part of COL
NRC	10 CFR Part 30	Byproduct License	Approval to possess fuel and source material.	To be issued as part of COL
NRC/U.S. Environmental Protection Agency	Resource Conservation and Recovery Act (RCRA), Atomic Energy Act, 40 CFR Part 266	Low-Level Mixed Waste Conditional Exemption	Allows the storage and treatment of low-level mixed waste.	Not yet submitted
Department of Energy (DOE)	Nuclear Waste Policy Act (42 USC 10101 et seq.) and 10 CFR Part 961	Spent Fuel Contract	The DOE Standard Contract for disposal of spent nuclear fuel contained in 10 CFR Part 961.	DE-CR01-11GC1126
Federal Aviation Administration (FAA)	14 CFR 77.13, Federal Aviation Act	Notice of Proposed Construction or Alteration	Construction of structures (>200 ft) affecting air navigation.	Not yet submitted
Department of Transportation (DOT)	49 CFR Part 107, Subpart G	Hazardous Materials Certificate of Registration	Shipment of radioactive and hazardous materials.	Reg. No. 061009 551 033RT[c]
U.S. Coast Guard	14 USC 81, 83, 85, 633 33 CFR Part 66	Authorization to Impact Navigation/Private Aids to Navigation	The interference of existing navigation aids or the placement and use of private aids to navigation in navigable waters of the United States.	Not yet submitted

Table H-1. (contd)

Agency	Authority	Requirement	Activity Covered	Status[b]
U.S. Army Corps of Engineers (USACE)	Section 10 of the Rivers and Harbors Appropriation Act of 1899, 33 USC 403 et seq.	Section 10 Permit	Structures and/or work that may affect navigability of any navigable waters of the United States. Structural alterations may include barge slip construction and the installation of or modification to existing intake and outfall structures.	Included in Joint Permit Application submitted to USACE on September 9, 2011
USACE	Clear Water Act (CWA), Section 404, 33 USC 1344	Section 404 Permit	Discharge of dredge or fill material within waters of the United States, including wetlands.	Included in Joint Permit Application submitted to USACE on September 9, 2011
U.S. Fish and Wildlife Service (FWS)	Endangered Species Act (ESA) Section 7, 16 USC 1539	ESA Section 7 Consultation	Consultation regarding the potential impacts on Federally threatened and endangered species.	Biological Assessment submitted March 30, 2012; concurrence from FWS on June 8, 2012.
FWS	Bald and Golden Eagle Protection Act (BGEPA), 16 USC 668	BGEPA Consultation	Consultation regarding the potential impacts on bald and golden eagles.	Ongoing
FWS	Migratory Bird Treaty Act (MBTA), 16 USC 703	MBTA Consultation	Consultation regarding the potential impacts on protected migratory birds.	Ongoing
State Authorizations				
Michigan Department of Environmental Quality (MDEQ) Office of Great Lakes	CZMA, 16 USC 1451 et seq.	Coastal Zone Management Act (CZMA) consistency review	Obtaining a Federal license or permit.	Issuance of Permit Number 10-58-0011-P (January 24, 2012) provides CZMA consistency determination

Table H-1. (contd)

Agency	Authority	Requirement	Activity Covered	Status[b]
MDEQ Water Resources Division	MCL 324.30306 *et seq.*; CWA, Section 404, 33 USC 1344	Wetlands Protection Permit	Any projects on or in wetlands regulated by the State of Michigan.	Permit Number 10-58-0011-P issued January 24, 2012
MDEQ Water Resources Division	MCL 324.32501 *et seq.*	Great Lakes Submerged Lands Permit	Dredging, filling, modifying, constructing, enlarging, or extending of structures in Great Lakes waters or below the ordinary high water mark of the Great Lakes; or connecting any natural or artificial waterway, canal, or ditch with any Great Lake including Lake St. Clair.	Permit Number 10-58-0011-P issued January 24, 2012

Permit Number 11-58-0055-P was issued April 25, 2012 (maintenance dredging) |
MDEQ Water Resources Division	MCL 324.32723	Water Withdrawal Permit	Withdrawals from the Great Lakes and connecting waterways of over 5 MGD.	Not yet submitted
MDEQ Water Resources Division	MCL 324.32705	Water Withdrawal Registration	Development of the withdrawal capacity on the property of an additional 100,000 gal of water per day from the waters of the State.	Not yet submitted
MDEQ Water Resources Division	MCL 324.4101 *et seq.*	Wastewater Facilities Construction Permit/Part 41 Construction Permit	Construction or modification of sewers, pumping stations, force mains, and treatment plants.	Not yet submitted
MDEQ Water Resources Division	33 USC 1251 *et seq.*

MCL 324.3101 *et seq.*

MCL 324.3301 *et seq.* | National Pollutant Discharge Elimination System (NPDES) Permit | Discharge of waste, waste effluent, and certain categories of stormwater runoff into the surface waters of Michigan during operation of the facility. | Permit Number MI0058892 issued February 2, 2012 |
| MDEQ Water Resources Division | MCL R323.2190 | NPDES Permits, Stormwater Construction Permit | A Permit by Rule may be obtained to authorize stormwater discharges from a construction site greater than or equal to 5 ac. | Not yet submitted |

Table H-1. (contd)

Agency	Authority	Requirement	Activity Covered	Status[b]
MDEQ Water Resources Division	33 USC 1251 et seq. MCL 324.3101 et seq.	NPDES General Dredging Dewatering Water Permit	Discharges of dredging dewatering water resulting from the removal of uncontaminated sediment from a waterway.	General Permit Number MIG690000[c]
MDEQ Water Resources Division	33 USC 1251 et seq. MCL 324.3101 et seq.	NPDES General Hydrostatic Pressure Test Water	Discharges from the hydrostatic pressure testing of new and existing piping, tanks, vessels, and other associated equipment that have been physically cleaned and/or provided with effluent treatment.	Permit Number MIG6790000[c]
MDEQ Water Resources Division	CWA Section 401, 33 USC 1341	Section 401 Water Quality Certification	The construction and operation of a facility that may result in any discharge into the navigable waters that will require a Federal license or permit.	The Wetlands Protection Permit (January 24, 2012) for construction and the NPDES Permit (February 2, 2012) for operation provide the Section 401 Water Quality Certification
MDEQ Resource Management Division	MCL R299.9303 et seq.	Hazardous Waste Management, Site Identification Number	A generator shall not treat, store, dispose of, or transport or offer for transport hazardous waste without having received a site identification number from the regional administrator.	Permit Number MID 087 056 685[c]
MDEQ Resource Management Division	MCL 29.5c	Review, Approval, and Certification of Aboveground Storage Tank (AST) Systems	Regulation of installation of new AST systems with individual tanks having a storage capacity of more than 1100 gal of flammable liquid or combustible liquid.	Not yet submitted

Table H-1. (contd)

Agency	Authority	Requirement	Activity Covered	Status[b]
MDEQ Resource Management Division	MCL R299.9822	Low-Level Mixed Waste Conditional Exemption	Low-level mixed waste storage and treatment conditional exemption eligibility and standards.	Not yet submitted
MDEQ Resource Management Division	MCL 333.13505	Radioactive Material Registration	Possession of radioactive materials.	Not yet submitted
MDEQ Air Quality Division	The Natural Resources and Environmental Protection Act (NREPA), Public Act 451 of 1994, as amended, Part 55 (Air Pollution Control) MCL R336.1201	Permit to Install	Construction of any air emission source.	Not yet submitted
MDEQ Air Quality Division	NREPA Part 55 (Air Pollution Control) MCL R336.1210-R336.1218 40 CFR Part 70	Air Permit	Operation of a source of air pollutants.	Not yet submitted
Michigan State Historic Preservation Office (SHPO)	National Historic Preservation Act of 1966, as amended (NHPA), Section 106, 36 CFR Part 800	Consultation with Michigan State Historic Preservation Office (SHPO), Federally recognized Indian Tribes, and other consulting parties	Consultation concerning the potential impacts on cultural resources.	Memorandum of Agreement executed March 20, 2012
Michigan Department of Natural Resources (MDNR)	MCL 324.36501 et seq.	Endangered Species Permit	Taking or harming of State-listed endangered species.	Not yet submitted

Table H-1. (contd)

Agency	Authority	Requirement	Activity Covered	Status[b]
MDNR	MCL 324.36501 *et seq.*	Consultation	Consultation regarding the potential impacts on threatened and endangered species.	Ongoing
Michigan Department of Transportation (MDOT)	MCL 259.481 *et seq.*	Tall Structures Act Permit	Construction of an object that has the potential to affect navigable airspace (height in excess of 200 ft or within 20,000 ft of an airport).	Not yet submitted
MDOT	MCL 247.171 *et seq.*	Construction Permits (Right–of-Way [ROW] Permit)	Activities by businesses or private parties and utility companies wishing to use the highway ROW for operations other than normal vehicular or pedestrian travel are required to obtain a permit from MDOT.	Not yet submitted
MDOT	MCL 257.716 *et seq.*	Transport permit	Movement over state highways of vehicles or loads that exceed the size or weight limitations specified by law.	Not yet submitted
Michigan Department of Community Health	MCL 333.13522	X-Ray Equipment Registration	Possession of a radiation machine.	Not yet submitted
Local Authorizations				
City of Monroe, Michigan	33 USC 1251 *et seq.* Michigan Water Resource Act Codified Ordinances of Monroe, Michigan, Streets, Utilities and Public Services Code, Chapter 1042, Division 2 Section 1042.15	Monroe Metropolitan Water Pollution Control Facility Industrial Pretreatment Permit	Treatment of wastewater to comply with categorical pretreatment standards and local limits.	Permit No. 1020[c]

Table H-1. (contd)

Agency	Authority	Requirement	Activity Covered	Status[b]
City of Monroe, Michigan/ Frenchtown Township	Codified Ordinances of Monroe, Michigan, Streets, Utilities and Public Services Code, Chapter 1042, Division 15 Section 1042.71	Sanitary Sewer Service Connection Permit	Required before a person uncovers, makes any connection with or opening into, uses, alters, or disturbs any public sewer or appurtenance to.	Not yet submitted
Frenchtown Township	Frenchtown Charter Township Zoning Ordinance No. 200 Article 6, Section 6.04 and Article 27.00, Section 27.06	Site Plan and Development Approval	Review of planned construction activities. Requires submittal of application for Site Plan Approval, which requires review of items such as engineering. The approval process may also result in the issuance of permits such as a grading permit issued under the authority of the Building Official.	Not yet submitted
Frenchtown Township		Engineering Review	Review of detailed engineering construction plans addressing water, sanitary, stormwater drainage, grading, and paving for the site.	Not yet submitted
Frenchtown Township	Frenchtown Charter Township Zoning Ordinance No. 200	Occupancy Permit	Occupancy of the building.	Not yet submitted
Frenchtown Township	Frenchtown Charter Township Zoning Ordinance No. 200 Article 4, Section 4.40 and Article 24, Section 24.05	Building Permit	Permit authorizing the construction, removal, moving, alteration, or use of a building or construction of any driveway or parking lot constructed of hard surface materials.	Not yet submitted
Frenchtown Township	Frenchtown Charter Township Zoning Ordinance No. 200 Article 20	Special Approval of Activities within Either the Floodway or Floodway Fringe	Approval of activities within the floodway area of floodway, fringe area of the floodway, or floodplain district.	Not yet submitted

Table H-1. (contd)

Agency	Authority	Requirement	Activity Covered	Status[b]
Frenchtown Township	Frenchtown Charter Township Zoning Ordinance No. 200 Article 4, Section 4.10	Temporary Building Used During Construction	Use of a portable structure as a temporary building during construction.	Not yet submitted
Frenchtown Township	Frenchtown Charter Township Zoning Ordinance No. 200 Article 26, Section 26.04	Landscape Development Plan	Submittal of a landscape development plan that illustrates areas of existing trees or wood lots that will be removed and those that will be retained.	Not yet submitted
Frenchtown Township	Frenchtown Charter Township Zoning Ordinance No. 200 Article 4, Section 4.21.2	Excavation Permit	Activities that propose to fill an area of 20,000 ft^2 or greater or any excavation and removal regardless of area involved except for mineral mining operations, farm ponds, and landscape ponds.	Not yet submitted
Monroe County, Michigan, Office of On-site Water Supply/Frenchtown Township	Codified Ordinances of Monroe, Michigan, Monroe County Environmental Health/Sanitary Code, Chapter III – Water Supplies	Well Permit	Construction of water supply wells, irrigation wells, heat exchange wells, industrial wells for water supply, test wells to obtain information regarding groundwater quantity or quality, recharge well, dewatering well, fresh water well at oil or gas well-drilling site.	Not yet submitted
Monroe County, Michigan, Drain Commissioner	Local Ordinance	Engineering Review	Review of surface water flow during operation.	Not yet submitted
Monroe County, Michigan, Drain Commissioner	NREPA Part 91, of Act 451 of the Michigan Public Acts of 1994 MCL 324.9101 *et seq.*	Soil Erosion and Sedimentation Control (SESC) Permit	Any earth change that disturbs 1 or more acres or is within 500 ft of a lake or stream.	Not yet submitted

Table H-1. (contd)

Agency	Authority	Requirement	Activity Covered	Status[b]
Monroe County, Michigan, Drain Commissioner	Act No. 40 of 1956	Drain Culvert Permit	Permit to construct in a drain.	Not yet submitted
Monroe County, Michigan, Health Department/ Frenchtown Township	Monroe County Environmental Health/ Sanitary Code, Chapter III, Section 302 Part 127 of Michigan Public Health Code, 1978 PA 368, as amended	Water Supply Permit	Any new construction or extensive change affecting the basic unit or the suction line on any water supply system within Monroe County, Michigan.	Not yet submitted

(a) Federal, State, and local authorizations that are required for building or operational activities are included. There are no Native American tribes with jurisdictional authority over activities at the Fermi site.

(b) Detroit Edison states in the ER that all necessary permits will be applied for in a timely manner. New permits may not be obtained in certain instances due to potential authorization of construction and operational activities through the modification of existing permits possessed by the Fermi Station.

(c) Permit authorizing current activities associated with operations on the Fermi site. When practical, existing permits will be modified to authorize activities associated with the construction or operation of a new nuclear facility on site.

Appendix I

Severe Accident Mitigation Alternatives

Appendix I

Severe Accident Mitigation Alternatives

I.1   Introduction

The Detroit Edison Company (Detroit Edison) has submitted an application to construct a General Electric-Hitachi Nuclear Energy, LLC- (GEH-) designed Economic Simplified Boiling Water Reactor (ESBWR) at the Enrico Fermi Atomic Power Plant (Fermi) site. Current policy developed after the Limerick decision (Limerick Ecology Action vs. NRC 1989) requires that the U.S. Nuclear Regulatory Commission (NRC) staff consider alternatives to mitigate the consequences of severe accidents in a site-specific environmental impact statement (EIS). The severe accident mitigation alternative (SAMA) review presented here considers both severe accident mitigation design alternatives (SAMDAs) and procedural alternatives.

In Title 10 of the Code of Federal Regulations (CFR), specifically 10 CFR 52.79(a)(38), the NRC requires that applicants for combined licenses (COLs) include "a description and analysis of design features for the prevention and mitigation of severe accidents" in the Final Safety Analysis Report (FSAR). Detroit Edison provides this information in Part 2 of the COL application. The Environmental Report (ER) (Part 3 of the COL application) also includes information regarding the SAMA analysis (Detroit Edison 2011).

In 10 CFR 52.47(a)(23), the NRC requires that applications for a reactor design certification include "a description and analysis of design features for the prevention and mitigation of severe accidents...." In addition, 10 CFR 52.47(a)(27) requires a description of a "plant-specific probabilistic risk assessment (PRA) and its results," and in 10 CFR 52.47(b)(2) the NRC requires an Environmental Report (ER) that contains the information required by 10 CFR 51.55. GEH has submitted all this information in documents that are part of the application for certification of the ESBWR design. Specifically, GEH has provided technical documents covering Revision 6 of the ESBWR PRA (GEH 2010a) and Revision 4 of the ESBWR SAMDA (GEH 2010b).

The NRC staff conducted a review of the Detroit Edison SAMDA analysis specific to operation of an ESBWR at the Fermi site. The staff reviewed the input parameters and values used by Detroit Edison (Detroit Edison 2011) for appropriateness, including information prepared by GEH in support of the ESBWR design certification. The Detroit Edison analysis is based on (1) the Revision 4 PRA (GEH 2009) and SAMDA analysis (GEH 2007) for the ESBWR design certification, and (2) results of the analysis of probability-weighted risks of the ESBWR design at the Fermi site described in Section 5.11.2 of this EIS.

An analysis for an ESBWR at a generic site is presented first, and then the analysis is extended to include consideration of Fermi site-specific information. These analyses have been updated by the NRC staff based on ESBWR PRA Revision 6 (GEH 2010a). The SAMDA analysis for the proposed ESBWR design certification has been reviewed and accepted by the staff as part of the design certification process (76 FR 14437).

I.2 ESBWR SAMDA Review – Generic Site

This section addresses the generic analysis of SAMDAs conducted by GEH, the applicant for certification of the ESBWR design. The SAMA review in Section I.3 extends the generic SAMDA analysis to include Fermi site-specific factors including meteorology, population, and land use. Section I.3 also addresses SAMAs that were not included in the generic analysis because they do not involve reactor system design.

I.2.1 ESBWR PRA and Consequence Results

GEH, the applicant for certification of the ESBWR design, conducted Level 1, Level 2, and Level 3 PRAs to estimate the core damage frequencies (CDFs) and offsite risk consequences that might result from a large number of initiating events and accident sequences. Table I-1 lists these CDF estimates and estimates of the large release frequencies (LRFs). Releases other than technical specification limits, when the containment is intact, are considered to be large. Table I-1 also lists NRC staff goals related to CDFs and LRFs.

Although this table does not provide quantitative estimates of CDFs and LRFs for fire, flood, and high-wind events during shutdown, they are discussed in ESBWR PRA Chapter 17 (GEH 2010a). Chapter 15 of the ESBWR PRA presents the results of a seismic margins analysis in which PRA methods are used to identify potential vulnerabilities in the design and so corrective measures can be taken to reduce risk. Based on the design considerations, risks associated with the seismic events are considered to be insignificant by GEH.

Chapter 10 of the ESBWR PRA Revision 6 (GEH 2010a) of the design certification application for the ESBWR design provides the results of Level 3 PRA in terms of an estimate of the offsite risk to the population within a 10-mi radius of a generic ESBWR location with conservative siting characteristics. The baseline results of the PRA for internal events during full-power operation are presented and compared to the Commission's individual and societal safety goals in Table I-2.

Table I-1. Comparison of ESBWR PRA Results with the Design Goals

Event Type	NRC Design Goal[a]		ESBWR PRA Results[b]	
	Core Damage Frequency (per Ryr)	Large Release Frequency (per Ryr)	Core Damage Frequency (per Ryr)	Large Release Frequency (per Ryr)
Internal at-power events	1.0×10^{-4}	1.0×10^{-6}	1.7×10^{-8}	1.4×10^{-9}
At-power internal flood events	1.0×10^{-4}	1.0×10^{-6}	7.0×10^{-9}	4.1×10^{-9}
At-power fire events	1.0×10^{-4}	1.0×10^{-6}	1.3×10^{-8}	1.6×10^{-9}
At-power high-wind events	1.0×10^{-4}	1.0×10^{-6}	8.5×10^{-8}	1.2×10^{-9}
Internal shutdown events	1.0×10^{-4}	1.0×10^{-6}	1.7×10^{-8}	1.7×10^{-8}

(a) SECY-90-016 (NRC 1990).

(b) From Chapter 17 of the ESBWR PRA Revision 6 (GEH 2010a).

Table I-2. Comparison of ESBWR PRA Results for a Generic Site with the Commission's Safety Goals

Goal	Risk Goal	ESBWR 24 hours after Onset of Core Damage (ground release)	ESBWR 72 hours after Onset of Core Damage (elevated release)	Safety Goal Achieved 72 hours after the Onset of Core Damage
Individual risk (0–1 mi)	$<3.9 \times 10^{-7}$ (0.1%)	1.6×10^{-10}	1.6×10^{-10}	Yes
Societal risk (0–10 mi)	$<1.7 \times 10^{-6}$ (0.1%)	2.0×10^{-11}	2.6×10^{-11}	Yes
Radiation dose[a] probability at 0.25 Sv (0–0.5 mi)	$<10^{-6}$	2×10^{-9}	2×10^{-9}	Yes

Source: Table 10.4-2 of GEH 2010a

(a) The values listed are radiation dose probability at 0.20 Sv, which is more bounding.

These results indicate that the risk from severe accidents would be at least four orders of magnitude lower than the Commission's safety goals (51 FR 30028).

The ESBWR PRA Revision 6 includes values for all external events and shutdown modes except for seismic events. Table 10.4.2 of the ESBWR PRA provides results for the external event and shutdown modes similar to those presented in Table I-2. For example, the total individual risk from internal and external events, 24 hours after onset of core damage, at both power and shutdown, is approximately 1.8×10^{-8}, which is less than the risk goal.

I.2.2   Potential Design Improvements

In the ER submitted as part of the design certification application (GEH 2010b), GEH identified 177 candidate alternatives based on a review of alternatives for other plant designs, including those considered in license renewal environmental reports and in the General Electric Advanced Boiling-Water Reactor (ABWR) SAMDA study (GE 1994), and on consideration of plant-specific enhancements. The candidate alternatives were then screened to identify candidates for detailed evaluation. The categories used in screening were as follows:

- Not applicable

- Already incorporated into the ESBWR design

- Not a design alternative (not required for design certification)

- Alternative prevention or mitigation functions extant

- Very low benefit

- Excessive implementation cost

- Consideration for further evaluation.

The development of the ESBWR design has benefitted from insights gained in numerous PRAs. The low CDFs and LRFs in Table I-1 are attributable to the implementation of improvements already incorporated into the design. The following are examples of enhancement features currently included in the ESBWR design:

- Improved isolation condenser system design

- Depressurization valves

- Alternating current (AC) independent fire water pumps for makeup and injection

- Passive containment cooling system

- Basemat internal melt arrest and coolability device and gravity-driven cooling system deluge function

- Direct current (DC) power reliability

- Actuation logic reliability

- Motor-driven, feed-water pumps

- Water pool elevation above drywell head elevation

- Containment ultimate strength and maximum design pressure

- Incorporation of flood mitigation into design

- Reactor water cleanup system heat exchanger sized for decay heat removal

- 72-hr coping period for station blackout

- Upgraded low-pressure piping for the reactor coolant pressure boundary

- Digital instrumentation and control systems.

The screening process eliminated 40 candidate alternatives as being not applicable to the ESBWR design; 71 candidate alternatives were considered to be similar to those already included in the ESBWR design, and 27 candidate alternatives were identified as procedural or administrative rather than design alternatives (whose benefits were considered to be unlikely to exceed those alternatives evaluated relative to their potentially high costs). Of the remaining 39 candidate alternatives, 37 were ruled out for cases in which other design features already perform the proposed function or obviate its need, and 2 were considered to have very low benefit because their insignificant contribution to reducing risk did not outweigh their excessive implementation costs. No candidate alternatives were identified for further evaluation.

I.2.3   Cost-Benefit Comparison

GEH used the cost-benefit methodology guidance in NUREG/BR-0184, *Regulatory Analysis Technical Evaluation Handbook* (NRC 1997), to calculate the maximum attainable benefit associated with completely eliminating all risk for the ESBWR.

This methodology involves determining the net value for a SAMDA according to the following formula:

$$Net\ Value = (APE + AOC + AOE + AOSC) - COE$$

where:

APE = present value of averted public exposure ($)
AOC = present value of averted offsite property damage costs ($)
AOE = present value of averted occupational exposure costs ($)
AOSC = present value of averted onsite costs ($); this includes cleanup, decontamination, and long-term replacement power costs
COE = cost of enhancement ($).

If the net value of a SAMDA is negative, the cost of implementing the SAMDA is larger than the benefit associated with the SAMDA, and it is not considered to be cost-beneficial.

To assess the risk reduction potential for SAMDAs, GEH assumed that each design alternative would work perfectly to completely eliminate all severe accident risk from the events that were evaluated. This assumption is conservative because it maximizes the benefit of each design alternative. GEH estimated the public exposure benefits for the design alternative on the basis of the reduction of risk expressed in terms of whole body person-rem per year received by the total population within a 50-mi radius of the generic ESBWR site.

Table I-3 summarizes the GEH's and NRC staff's estimates of each of the associated cost elements. The results are based on the approach, parameters, and data listed in NUREG/BR-0184. GEH's estimates in Table I-3 are based on the PRA Revision 5 CDF of 1.12×10^{-7} per reactor-year (Ryr) (GEH 2010c), which are similar to those in PRA Revision 6 (GEH 2010a). (The total CDF in the Revision 4 PRA is 1.2×10^{-7} per Ryr [GEH 2009].) The CDF is driven by high core damage frequencies from internal and high-wind events during shutdown. GEH used the results from the ESBWR Level 3 PRA, namely, an offsite population dose risk of 0.035 Sv/Ryr and an offsite cost risk of $1931/Ryr based on input from the Electric Power Research Institute Advanced Light Water Reactor Utility Requirement Document (GEH 2010c).

GEH provided the present value estimates for the various attributes using a 3 percent discount rate and the maximum parameter values provided in NUREG/BR-0184. Revision 4 of NUREG/BR-0058, *Regulatory Analysis Guidelines of the U.S. Nuclear Regulatory Commission* (NRC 2004), reflects the agency's policy on discount rates. NUREG/BR-0058 Revision 4 states that two sets of estimates should be developed: one at 7 percent and one at 3 percent for sensitivity analysis.

The monetary present value estimate for each risk attribute does not represent the expected reduction in risk resulting from a single accident; rather, it is the present value of a stream of potential losses extending over the projected lifetime of the facility (in this case, projected to be 60 years). Therefore, the estimate reflects the expected annual loss resulting from a single accident, the possibility that such an accident could occur at any time over the licensed life, and the effect of discounting these potential future losses to present value.

GEH estimated the total present dollar value equivalent associated with complete elimination of severe accidents at a single ESBWR unit site to be $397,863 (see Table I-3 below). Therefore, for any SAMDA to be cost-beneficial, the enhancement cost must be less than $397,863. GEH assessed the capital cost associated with two design alternatives evaluated for the ESBWR. For both design alternatives, GEH stated that the implementation cost would be more than $1 million (GEH 2010b). Based on the averted cost estimate of $397,863, GEH concluded that none of the SAMDA candidates are cost-beneficial, because any design change costs would far exceed this value.

Table I-3. Summary of Estimated Averted Costs for a Generic Site

Quantitative Attributes		Present Value Estimate ($)		
		NRC Staff Best Estimate[a]	GEH Maximum[b]	NRC Staff Maximum[c]
Health	Public	100,000[d]	194,740	197,720[d]
	Occupational	56	249	250
Property	Offsite	27,200[d]	53,720	53,770[d]
	Onsite	NA[e]	NA	NA
Cleanup and decontamination	Onsite	1710	4674	4060
Replacement power		4520	144,480	148,020
Total		**133,486**	**397,863**	**403,820**

Source: GEH 2010b

(a) "Best estimate" is based on mean release frequency (from Revision 5 of the PRA), "best estimate" parameter values in NUREG/BR-0184, and 7 percent discount rate.

(b) Maximum estimate is based on mean release frequency (from Revision 5 of the PRA), high or upper estimate parameter values in NUREG/BR-0184, and 3 percent discount rate.

(c) NRC staff maximum is based on parameter values used in (b), and release frequency from Revision 5 of the PRA.

(d) Estimated using the applicant-provided Electric Power Research Institute Advanced Light Water Reactor Utility Requirement Document, property damage, and the new release category frequencies (GEH 2010a).

(e) NA = Not analyzed.

Note: PRA Revision 5 release frequencies are the same as those in PRA Revision 6.

I.2.4 Staff Evaluation

In 10 CFR 52.47(a)(27), the NRC requires that an applicant for design certification perform either a plant-specific or site-specific PRA. The aim of this PRA is to seek improvements in the reliability of core and containment heat removal systems that are significant and practical. The set of potential design improvements considered for the ESBWR includes those from generic boiling water reactor SAMA reports and from the ABWR design. The ESBWR design already incorporates many design enhancements related to severe accident mitigation. Such design improvements have resulted in a CDF that is about an order of magnitude less than that of the ABWR design. For example, the ESBWR design can cope with a station blackout (SBO) for 72 hr (i.e., no reliance on AC power for the first 72 hr), thus eliminating CDF sequences that contributed more than 40 percent of CDF in the ABWR design.

GEH's risk reduction estimates are based on mean values of release frequencies and maximum-estimate parameter values from NUREG/BR–0184, without consideration of uncertainties in CDF or offsite consequences. Even though this approach is consistent with that used in previous design alternative evaluations, further consideration of these factors could lead to significantly higher risk reduction values, given the extremely small CDF and risk estimates in the baseline PRA. The uncertainties in CDF or in offsite radiation exposures are fairly large because key safety features of the ESBWR design are unique, and their reliability has been evaluated through analysis and testing programs rather than through operating experience.

The NRC staff's analyses of the total present value using the mean CDF and release frequencies from Revision 6 of the PRA and a 3 percent discount rate indicate a maximum value of about $403,820. NRC staff notes that the estimated averted public exposure is a major contributor. This arises from high release frequencies for internal and high-wind events during shutdown. For events during shutdown, the analysis conservatively assumes that core damage scenarios will lead to large releases. This is because, the containment is open during most of the shutdown period.

The second major contributor to the present value estimate is replacement power costs. The replacement power cost parameters recommended in NUREG/BR–0184 are based on a generic reactor operating at an average capacity factor of about 65 percent and on replacement energy costs in 1993 dollars, The total present dollar value would be even higher if the annual replacement power cost was adjusted for a future energy cost increase and the capacity factor was increased to 90 percent, which is the design operating assumption for the ESBWR. However, GEH used a very conservative approach in estimating the replacement power cost. GEH selected the parameter that corresponds to the 3 percent discount rate for the net present value of replacement power for a single event recommended in NUREG/BR-0184. Then GEH used this parameter as an input and estimated a new, more conservative net present value of the replacement power for a single event. This approach resulted in a net present value of replacement power that is about a factor of ten higher than the value estimated in NUREG/BR-0184. Even with this increase, which is more than what it would be if adjustments for the future energy cost increase and capacity factor were to be made, the present value estimate is still lower than the GEH's $1 million minimum cost estimate for a SAMDA. Also, the ESBWR CDF is very low on an absolute scale as compared to those of currently operating plants. Moreover, in view of the features already incorporated in the ESBWR design and the margin between the cost of SAMDAs evaluated and their potential benefits, any increase in benefits due to increased replacement power costs would not be significant enough to cause any SAMDAs to become cost-beneficial. Therefore, the NRC staff concludes that further evaluation of future energy cost and capacity factor increases is not warranted.

GEH indicated that any of the potential design modifications considered would cost a minimum of $1 million to implement, as indicated above. NRC staff considers the assertion of potential costs for the ESBWR acceptable, because it is reasonable to conclude that the cost of implementing (design, procurement, installation, testing, etc.) the design alternatives that were considered, such as constructing a building connected to the containment building or installing limit switches on all containment isolation valves, would far exceed GEH's $1 million minimum cost estimate. Therefore, a minimum cost of $1 million is approximately 2.5 times the maximum benefit of $403,820. The NRC staff concludes that no single modification would eliminate the total CDF and that none of the potential design modifications could be justified on the basis of cost-benefit considerations.

I.3   Fermi Site-Specific SAMDA Review

The discussion above evaluates SAMDAs for the ESBWR at a generic site. The following discussion updates that evaluation to include consideration of Fermi site-specific factors, including meteorological conditions, population distribution, and land use. It also updates the evaluation to include the results and the approach in PRA Revision 4 for the generic design. The last part of this discussion deals with SAMAs for procedures and training.

I.3.1   Risk Estimates

NRC staff evaluated the potential risks associated with severe accidents for an ESBWR by using Fermi site-specific data. Detroit Edison provided a site-specific consequence analysis using the Revision 4 PRA CDF (Detroit Edison 2011). Table 5-32 of this EIS, gives a population dose and a cost risk of 0.032 person-rem/Ryr and $110/Ryr, respectively, for the at-power internal events with a CDF of 1.7×10^{-8} per Ryr. The total environmental risk associated with both shutdown and power operations, including consideration of internal events, fires, high winds, and floods, is provided in Table 5-33 of this EIS, which gives a total population dose and a cost risk of about 2.3 person-rem/Ryr and $4900/Ryr, respectively.

I.3.2   Cost-Benefit Comparison

In Section 7.3.2 of the ER (Detroit Edison 2011), Detroit Edison estimates the averted costs associated with eliminating all severe accident risks for an ESBWR at the Fermi site. The analysis is an update of the GEH SAMDA analysis (GEH 2007) to include site-specific information. Detroit Edison substituted population dose and offsite cost risks based on 2060 population projections for the Fermi site for the population dose and offsite property costs in the GEH analysis.

Detroit Edison provided a site-specific cost-benefit analysis using the Revision 4 PRA CDF (Detroit Edison 2011). Detroit Edison provided an estimated total present dollar value equivalent associated with complete elimination of severe accidents at a single ESBWR unit site to range between $139,446 and $280,189 and concluded that no design changes would be cost-effective to implement (Detroit Edison 2011).

NRC staff evaluated the risk reduction potential of design improvements for the ESBWR at the Fermi site based on the Detroit Edison's risk reduction estimates for the various design alternatives, in conjunction with an assessment of the potential impact of uncertainties on the results. The staff performed the averted cost estimates with the parameters used by Detroit Edison and the upper bound values used in ESBWR SAMDA Revision 4 (GEH 2010b). The results of both the Detroit Edison and the NRC estimates of averted costs are presented in Table I-4. The NUREG/BR–0184 handbook provides two sets of parameters (best estimate and high estimate) for the parameters used in the calculations of the occupational dose after accident and during decontamination and cleanup, and for the replacement power costs. The NRC staff's maximum estimate is based on the use of "high or upper bound" estimated parameters in NUREG/BR-0184 and the ESBWR power rating of 1585 MW(e) that were used in ESBWR SAMDA Revision 4 (GEH 2010b). The major contributor to this estimate is the use of the GEH's high value for the long-term replacement power costs parameter for a 910-MWe "generic" reactor in NUREG/BR–0184. The use of the GEH's high value increases the replacement power costs by about a factor of 10 over the best estimate (see Table I-4, Columns 6 and 7). As stated in Section I.2.4, this increase replacement power cost is well above any potential change for adjustments in the future energy cost increase and capacity factor.

The NRC staff's analyses of the total present value using the mean CDF and release frequencies from Revision 6 of the PRA and a 3 percent discount rate indicate a maximum value of about $422,000. The NRC staff noted that any design modifications would be costly, and a single modification would not eliminate the total CDF. On the basis of results presented in Table I-4, the NRC staff agreed with Detroit Edison's conclusion that no design change would be cost-beneficial.

1.3.3 Procedural and Training SAMAs

The original list of 177 ESBWR SAMDAs included 27 candidate alternatives that were procedural or administrative in nature. These items were eliminated from consideration because they did not involve design changes. Examples of items removed from consideration for this reason are as follows:

- Enhance procedural guidance for use of cross-tied component cooling or service water pumps.

- Implement procedures for alignment of a spare diesel to shut down board after loss of offsite power and failure of diesel normally supplying it.

- Emphasize steps in recovery of offsite power after an SBO.

- Develop a severe weather conditions procedure.

- Develop procedures for replenishing diesel fuel.

Table I-4. Summary of Estimated Averted Costs for the Fermi Site

		Present Value Estimate ($)				
		Detroit Edison[a]		NRC Estimates[b]		
Quantitative Attributes		7% Discount	3% Discount	7% Discount	3% Discount	Maximum[c] Estimate
Health	Public	64,166	126,875	65,855	130,213	130,213
	Occupational	58	133	58	133	258[c]
Property	Offsite	68,950	136,335	68,247	134,943	134,943
	Onsite	NA[d]	NA	NA	NA	NA
Cleanup and decontamination	Onsite	1761	4184	1761	4183	4183
Replacement power		4512	12,668	4658	13,077	152,565[c]
Total		139,446	280,189	140,579	282,549	422,162

(a) From Fermi 3 Environmental Report support documentation, based on PRA Revision 4 CDF estimates.

(b) NRC staff estimates are based on PRA Revision 6 CDF estimates and Chapter 10, Table 10.3-3c, release category assignments (GEH 2010a), and the EBWR power rating of 1585 MW(e) used in ESBWR SAMDA Revision 4 (GEH 2010b). The ESBWR power rating in the Detroit Edison's analysis is 1535 MW(e). This change will affect only the replacement power cost.

(c) The maximum estimate is based on "high or upper bound" estimated parameters in NUREG/BR-0184 and the ESBWR power rating of 1585 MW(e) and on replacement power parameters in ESBWR SAMDA Revision 4 (GEH 2010b).

(d) NA = Not analyzed.

- Increase frequency for valve leak testing. Improve inspection of rubber expansion joints on the main condenser.

These candidate alternatives fall within the scope of the SAMA review that the NRC conducts as part of the environmental review of applications. However, such SAMAs generally involve operational and training procedures that have not been developed for a reactor and are typically not developed until construction has been completed and the plant is approaching operation.

The staff reviewed the candidate alternatives that were previously screened out because they did not involve design changes. Because the maximum attainable benefit is so low, a SAMA based on procedures or training for an ESBWR at the Fermi site would have to reduce the CDF or risk to near zero to become cost-beneficial. Based on its evaluation, the staff concludes that it is unlikely that any of the SAMAs based on procedures or training would reduce the CDF or risk that much. Therefore, the staff further concludes it is unlikely that these SAMAs would be cost-effective.

Detroit Edison states that it will consider the procedural and administrative SAMAs when it is developing its procedures, as long as they do not exceed the maximum averted cost. Detroit Edison makes this statement through a commitment (COM ER 7.3-002) which states (Detroit Edison 2011):

> SAMA analysis to comply with 40 CFR 1502.16(h) shall be conducted of the administrative and procedural measures applicable to Fermi 3 and considered for implementation prior to fuel load if the associated cost does not exceed the maximum value associated with averting all risk of severe accidents.

Based on this statement, the staff expects that Detroit Edison will consider risk insights and mitigation measures in the development of procedures and training; however, this expectation is not crucial to the staff's conclusions because the staff already concluded procedural and training SAMAs would be unlikely to be cost-effective.

I.4 Conclusions

Based on the evaluation of the ESBWR PRA (GEH 2010a) and SAMDA analysis (GEH 2010b), the Fermi site-specific severe accident and SAMDA analysis (Detroit Edison 2011), and its own independent review, the staff concludes that there are no ESBWR SAMDAs that would be cost beneficial at the Fermi site. The staff expects that Detroit Edison will use risk insights and mitigation measures in the development of procedures and training; however, this expectation is not crucial to the staff's conclusions because the staff already concludes procedural and training SAMAs would be unlikely to be cost-effective.

I.5 References

10 CFR Part 51. Code of Federal Regulations, Title 10, *Energy*, Part 51, "Environmental Protection Regulations for Domestic Licensing and Related Regulatory Functions."

10 CFR Part 52. Code of Federal Regulations, Title 10, *Energy*, Part 52, "Licenses, Certifications, and Approvals for Nuclear Power Plants."

40 CFR Part 1502. Code of Federal Regulations, Title 40, *Protection of Environment*, Part 1502, "Environmental Impact Statement."

51 FR 30028. August 21,1986. "Safety Goals for the Operation of Nuclear Power Plants; Policy Statement; Correction and Republication." *Federal Register.* U.S. Nuclear Regulatory Commission.

76 FR 14437. March 16, 2011. "Economic Simplified Boiling Water Reactor Standard Design: GE Hitachi Nuclear Energy; Issuance of Final Design Approval." *Federal Register.* U.S. Nuclear Regulatory Commission.

Detroit Edison Company (Detroit Edison). 2011. *Fermi 3 Combined License Application, Part 3 Environmental Report.* Revision 2, Detroit, Michigan. February. Accession No. ML110600498.

General Electric (GE). 1994. *Technical Support Document for the ABWR.* General Electric Company, San Jose California. December.

General Electric-Hitachi Nuclear Energy Americas, LLC (GEH). 2007. *Licensing Topical Report, ESBWR Severe Accident Mitigation Design Alternatives.* NEDO-33306, Revision 1, August 14. Accession No. ML072390051.

General Electric-Hitachi Nuclear Energy Americas, LLC (GEH). 2009. *ESBWR Certification Probabilistic Risk Assessment.* NEDO-33201, Revision 4. June 26. Accession No. ML092030199 and ML092030244 for Chapter 10.

General Electric-Hitachi Nuclear Energy Americas, LLC (GEH). 2010a. *ESBWR Certification Probabilistic Risk Assessment.* NEDO-33201, Revision 6. October 7. Accession No. ML102880535.

General Electric-Hitachi Nuclear Energy Americas, LLC (GEH). 2010b, *Licensing Topical Report, ESBWR Severe Accident Mitigation Design Alternatives.* NEDO-33306, Revision 4. October 25. Accession No. ML102990433.

Appendix I

General Electric-Hitachi Nuclear Energy Americas, LLC (GEH). 2010c. *ESBWR Certification Probabilistic Risk Assessment.* NEDO-33201, Revision 5. February. Accession No. ML100740286.

Limerick Ecology Action vs. NRC, 869 F.2d 719 (3rd Cir. 1989).

U.S. Nuclear Regulatory Commission (NRC). 1990. "Evolutionary LWR Certification Issues and Their Relationships to Current Regulatory Requirements," SECY 9-016, January.

U.S. Nuclear Regulatory Commission (NRC). 1997. *Regulatory Analysis Technical Evaluation Handbook, Final Report.* NUREG/BR-0184, Office of Nuclear Regulatory Research, Washington, D.C. January. Accession No. ML050190193.

U.S. Nuclear Regulatory Commission (NRC). 2004. *Regulatory Analysis Guidelines of the U.S. Nuclear Regulatory Commission.* NUREG/BR-0058, Washington, D.C. September. Accession No. ML042820192.

Appendix J

U.S. Army Corps of Engineers
Public Interest Review Factors and
Detroit Edison's Onsite Alternatives Analysis

Appendix J

U.S. Army Corps of Engineers
Public Interest Review Factors and
Detroit Edison's Onsite Alternatives Analysis

This appendix presents (1) a summary of the factors that are considered by the U.S. Army Corps of Engineers (USACE) in its public interest review of applications for a permit to perform regulated activities that would affect waters of the United States and (2) an onsite alternatives analysis prepared by Detroit Edison Company (Detroit Edison) to demonstrate that its proposed site layout chosen for the proposed new Enrico Fermi Unit 3 (Fermi 3) at the Enrico Fermi Atomic Power Plant (Fermi) site would minimize impacts to jurisdictional wetlands and waters of the United States. These topics are addressed in Sections J.1 and J.2 of this appendix, respectively.

J.1 Public Interest Review Factors

As set forth in Title 33 of the Code of Federal Regulations (CFR) Part 320, a public interest review must be completed prior to any Department of the Army (DA) permit decision by the USACE. The USACE decision on whether to grant or deny a permit is based, in part, on an evaluation of the probable impact of the proposed activity and its intended use on the public interest. This evaluation is referred to as the "public interest review." The public interest review requires a careful weighing of all relevant factors in a particular case. The specific weight of each factor is determined by its importance and relevance to the proposed project. Some public interest review factors may be given greater weight, while others may not be relevant or as important based on project characteristics. The USACE public notice (USACE 2011), the Draft EIS public comment process, DEIS public meetings, and the EIS public scoping process have been the primary methods used to solicit public comment on the project's effect on public interest factors. Full consideration and appropriate weight will be given to all comments, including those of Federal, State, and local agencies, and other experts on matters within their expertise. The benefits and detriments of a project are balanced by considering effects on such public interest factors as conservation, economics, aesthetics, general environmental concerns, wetlands, historic properties, fish and wildlife values, flood hazards, floodplain values, land use, navigation, shore erosion and accretion, recreation, water supply and conservation, water quality, energy needs, safety, food and fiber production, mineral needs, considerations of property ownership, and, in general, the needs and welfare of the people. The conditions, including compensatory mitigation, under which a proposal would be allowed to go forward, would be developed and incorporated within the public interest review process to the extent that such conditions are found to be appropriate and practicable by the USACE. However, only the measures required to confirm that the project is not contrary to the public interest may be required in this specific context. This required public interest review ensures that a USACE permit decision reflects the National concern for both protection and utilization of important resources. The public interest review described above can be found in 33 CFR 320.4 and will be completed by the USACE as part of its evaluation of the Fermi 3 proposal for a DA permit.

J.2 Detroit Edison's Onsite Alternatives Analysis and Proposed Least Environmentally Damaging Practicable Alternative (LEDPA)

Activities involving the discharge of dredged or fill material into waters of the United States, including wetlands, typically require authorization from the USACE under Section 404 of the CWA. The CWA Section 404(b)(1) Guidelines (40 CFR Part 230) (Guidelines) are the substantive criteria the USACE uses to determine a project activity's environmental impact on aquatic resources attributable to the discharge of dredged or fill material. Among other things, an applicant for a 404 permit must demonstrate to the USACE that proposed project-related dredge or fill activities satisfy the Guidelines and constitute the least environmentally damaging practicable alternative (LEDPA). An applicant would typically conduct analyses of the impacts of its proposed actions involving dredge or fill discharges into waters of the United States and of alternatives to avoid and minimize impacts to identify a proposed LEDPA that still allows accomplishment of the overall project purpose and demonstrates compliance with the Guidelines. As part of this process, an applicant would initially submit a conceptual plan to address the mitigation of any remaining unavoidable adverse impacts to aquatic resources that would still occur after all practicable avoidance and minimization measures were applied.

Based on guidance provided by the USACE regarding Guidelines compliance, Detroit Edison conducted an onsite alternatives analysis to identify a practicable alternative that would avoid and minimize adverse impacts to waters of the United States. This analysis includes Detroit Edison's proposed LEDPA and is included at the end of this appendix (Appendix J). USACE has not verified the adequacy of Detroit Edison's proposed LEDPA at this time. However, USACE is actively reviewing and coordinating with Detroit Edison regarding its proposed LEDPA. USACE could potentially identify additional practicable avoidance and/or minimization measures during its evaluation that could result in the USACE-identified LEDPA having fewer adverse impacts on waters of the United States than Detroit Edison's proposed LEDPA, as presented in its analysis. Any subsequent changes to the proposed site plan and/or activities as a consequence of the USACE-identified LEDPA would result in fewer adverse impacts on waters of the United States than identified in the Final EIS.

To offset the Detroit Edison-identified unavoidable adverse impacts on aquatic resources as a result of its proposed LEDPA, Detroit Edison initially proposed a conceptual mitigation strategy that was included in Appendix K of the Draft EIS. The USACE LRE-2008-00443-1-S11 public notice (USACE 2011) provided additional opportunity for public comment on Detroit Edison's proposed LEDPA and concept mitigation strategy. Detroit Edison subsequently refined its mitigation strategy, based on coordination with USACE, and produced the mitigation plan that is now contained in Appendix K of this Final EIS. Detroit Edison's mitigation plan proposes to compensate for the unavoidable loss of aquatic function on the Fermi site by reestablishing comparable aquatic functions at an offsite location at an average replacement ratio of 3:1. The evaluation of alternative energy sources (e.g., power purchases, demand-side management,

fossil-fuel alternatives, and renewable energy alternatives), alternative sites (Fermi, Belle River–St. Clair, Greenwood, Petersburg, and South Britton), and system design alternatives (including heat dissipation and cooling system alternatives) are discussed in Chapter 9 of this EIS.

Section 4 of Detroit Edison's Joint Permit Application (Detroit Edison 2011), which presents their onsite alternatives analysis and proposed LEDPA determination, is provided in the remainder of this appendix.

Appendix J

SECTION 4: PROPOSED PROJECT PURPOSE, INTENDED USE, AND ALTERNATIVES CONSIDERED

1) Purpose/Intended Use:

Detroit Edison proposes to construct and operate a new nuclear power plant at the Fermi site. The proposed unit is to be designated as Fermi 3. The purpose of the Fermi 3 project is fourfold:

1. Generate a net electrical output of approximately 1,535±50 megawatts (MWe) for sale that will reliably aid in satisfying the forecasted energy and capacity needs of Detroit Edison customers located in the Detroit Edison Service Area;
2. Provide new baseload electric generation capacity as early as 2021 to compensate for the expected retirement of existing, aging baseload generating units and diminishing availability of the midwest independent service operator region's baseload generation capacity;
3. Provide price stability by minimizing reliance on imported power into the Detroit Edison service territory; and
4. Utilize an electric generation technology that is less subject to price fluctuations resulting from either fuel or regulatory drivers, provides fuel diversity, and reduces reliance on fossil fuel and their attendant environmental impacts.

The above purpose is in-line with Detroit Edison's mission to provide reliable and affordable electrical power.

Construction of a new nuclear electric generating facility is needed to provide reliable, affordable power to address Michigan's expected future peak electric demand. Detroit Edison has evaluated the need for power and the related benefits to be generated by the proposed facility. The need for power was assessed by balancing the current and forecasted demand against the current and forecasted supply, while demonstrating that an adequate reserve margin is maintained. Detroit Edison's assessment considered information regarding factors such as marketing, location, and history that influence or constrain the nature, size, price, and class of the project.

The need for power assessment is derived from the "Michigan 21st Century Electric Energy Plan" (Plan).[1] The Plan was prepared and issued by the Michigan Public Service Commission pursuant to Executive Directive No. 2006-02. The Plan reached several significant conclusions, including the following:

- Michigan's peak electric demand is forecasted to grow at approximately 1.2 percent per year for the next 20 years;
- There is a need for additional electric generating resources in order to preserve electric reliability and provide affordable energy over the next 20 years. This modeling outcome is confirmed even in the presence of increased use of energy efficiency and renewable resources;
- The projected electric demand will not be satisfied through the expansion of transmission nor access to external markets; and
- There is need for regulated baseload capacity to prevent natural gas prices from driving up wholesale costs and market prices for an increasing number of hours each year.

The above conclusions were based upon key factors such as the current age of baseload units and newer electric generating units' reliance on natural gas. As indicated above, the Plan concluded that the state of Michigan has a current need for new baseload capacity and the need is projected to increase. Michigan's current baseload generating units are an average of more than 48 years old.

---

[1] See http://www.dleg.state.mi.us/mpsc/electric/capacity/energyplan/index.htm.

The average age of Detroit Edison's coal-fired generation units is 44 years old. The last new baseload plant in the state of Michigan began commercial operation more than 18 years ago. The assessment assumes that older, less efficient units, totaling 3,755 MW of capacity, will be retired by 2025.

Further, new baseload electric production is needed due to the fact that recently constructed electric generation units in Michigan have been limited to natural gas-fired facilities. Natural gas-fired units currently represent approximately 29 percent of Michigan's generating capacity. Dependence upon natural gas-fired units has exposed Michigan to volatile electricity prices driven by fluctuating fuel market prices.

Detroit Edison evaluated alternative means of meeting the baseload generation need. That analysis concluded that coal-fired or natural-gas fired generation provide reasonable alternatives to Fermi 3 for meeting the identified need for new baseload generation. However, after considering the potential environmental impacts associated with these alternative energy sources, Detroit Edison determined they would not be environmentally preferable to the proposed Fermi 3 nuclear power plant.

2) Alternatives Considered:

Detroit Edison sought to avoid and minimize impacts to waters of the United States, including wetlands, associated with the proposed Fermi 3 project by evaluating practicable alternatives that would fulfill the project's purpose and need. Detroit Edison's alternatives analysis included consideration of alternative locations for new nuclear electric production consistent with the purpose and need described above. After determining that the Fermi site was the practicable alternative project location that would result in the least potential impacts to aquatic resources, Detroit Edison considered site layout alternatives to minimize potential wetland impacts in terms of both quantity and quality. Both components of the alternatives analysis are summarized below. Detroit Edison's alternatives evaluation illustrates that the proposed use of the Fermi site is the least environmentally damaging practicable alternative (LEDPA) that fulfills the project's purpose and need. Detroit Edison has also proposed mitigation for the unavoidable impacts to waters of the United States.

a) Alternative Sites

Detroit Edison reviewed the eight candidate sites identified through the site selection process described in Section 9.3 of the Fermi 3 Combined License Application Environmental Report within the context of the CWA Section 404(b)(1) guidelines to identify a LEDPA site. The candidate sites were evaluated for practicability to construct and operate a nuclear generating facility. The sites that were found to be practicable were then evaluated for potential impacts on waters of the United States and adjacent wetlands to identify an environmentally preferable location.

The candidate sites included five greenfield sites, two existing fossil-fired sites, and one existing commercial nuclear site. The practicability assessment considered various technical, economic, safety, and environmental criteria that reflect the overall purpose of the project. The results of that evaluation are summarized in **Table 4-1**. Six sites (five greenfield sites and one existing fossil-fired site) that exhibited undesirable characteristics were judged to be impracticable as sites for locating a new nuclear plant and were excluded from further review. The two remaining candidate sites, the Greenwood Energy Center site and the Fermi site, were then evaluated for impacts on waters of the U.S. and adjacent wetlands.

Detroit Edison evaluated the potential wetland and stream impacts associated with construction of the nuclear generating facility and any required infrastructure such as transmission corridors and make-up water supply or blowdown discharge pipelines to support the closed-cycle cooling

Appendix J

system. The potential impacts associated with nuclear development at the Fermi and Greenwood sites are summarized in the **Table 4-2**. Based on the overall potential impacts to waters of the U.S., the Fermi site would be the LEDPA.

b)   Site Layout Alternatives

Detroit Edison proposes to construct and operate a new nuclear power plant at the Fermi site. The proposed unit is to be designated as Fermi 3. The Fermi site (the area within the Fermi property boundary) consists of approximately 1260 acres in eastern Monroe County, Michigan. The existing Fermi 2 unit is in the northeast part of the site. Fermi 3 and associated facilities will be located in an area south of the existing Fermi 2 protected area. Most of the land that will be occupied by Fermi 3 and associated facilities was disturbed during construction of Fermi 1 and Fermi 2; however, some construction will occur in areas that have been undisturbed for longer periods of time. This section discusses the onsite layout alternatives considered and the relevant impacts to aquatic resources associated with those alternatives for the Fermi 3 project.

The Fermi 3 site layout includes the power block, cooling tower, switchyard, parking, construction laydown areas, transmission lines, access road, cooling water intake structure, discharge pipe, and barge docking facility. Detroit Edison applied as much repositioning of project components as possible within project practicability limits to avoid and minimize impacts to wetlands and other natural resources at the Fermi site. Four project layout alternative scenarios were evaluated. These alternative layouts are identified as Revision 0, Revision 1, Revision 2, and the Preferred Alternative.

The site layout was evaluated for potential environmental impacts to the Fermi site. This analysis focused on environmental categories that are protected under special-purpose environmental laws and that contain specific provisions for the avoidance and minimization of impacts. These categories include wetlands, archaeological resources, and protected species. Complete avoidance of some impacts to environmental categories, such as wetlands, associated with Fermi 3 may not be feasible due to the large area of land disturbance required. Efforts were made to avoid impacts to wetlands through consideration of several different project alternatives.

A process to avoid, minimize, or compensate impacts to waters of the United States, including wetlands, was completed for the Fermi 3 project. This process included the consideration of alternative onsite locations for major structures and changes in site configuration to minimize damages to waters of the United States.

Key Constraints

Several key constraints guided the process of determining locations for Fermi 3 Nuclear Power Plant and construction-related activities relative to the available property on the Fermi site and the location and operational needs of the Fermi 2 Nuclear Power Plant. As this discussion will illustrate, unavoidable impacts to wetlands resulted when the key constraints could not be satisfied without incurring those temporary or permanent impacts.

The key constraints are as follows:

1)   The site layout must minimize impacts to the environment and to the Detroit River International Wildlife Refuge.
2)   Fermi 3 construction cannot interfere with the operations of the existing Fermi 2 Nuclear Power Plant.
3)   Fermi 3 construction cannot interfere with Fermi 2 security requirements or programs.
4)   Fermi 2 operations must not interfere with Fermi 3 construction.
5)   Fermi 2 operations must not interfere with federally mandated Fermi 3 security requirements, which are distinct from operating plant security requirements.

6) The location of the Fermi 3 power block must allow for both Fermi 2 and Fermi 3 plants to be combined into a single protected area security boundary after construction is completed that meets federally mandated security requirements. This will facilitate operational synergies such as sharing of personnel and common support facilities, the Primary Access Portal (PAP) to the protected area, warehouses, and maintenance shops.

7) The construction site must provide for a contiguous, unimpeded flow of personnel, equipment and materials.

8) The Fermi 3 construction site must have adequate, onsite space for the following: laydown and staging of materials; fabrication and assembly of modular components; and; construction support facilities. Nuclear power plant construction management consultants have advised Detroit Edison that a minimum of 100 acres of land should be available onsite, contiguous to or near the construction area, for these activities.

9) Placement of structures must satisfy nuclear safety requirements.

Constraint 1 has been a primary consideration throughout the site layout development process, however, as the project has moved forward, additional environmental studies and information have been developed which have been the principal driver for revisions to the proposed site layout to further minimize environmental impacts.

While the constraints have remained the same throughout the development of the site layout, as Detroit Edison's knowledge of site environmental conditions evolved, revised versions of the site layout were created in keeping with Constraint 1. Each of the four versions of the site layout satisfied the key constraints based upon the state of knowledge at the time the site revision was developed.

The method chosen to address Constraints 2 through 5 was to separate Fermi 2 operational activities from the Fermi 3 construction site the maximum extent. This separation resulted in Constraints 10 and 11, as follows:

10) All Fermi 2 operational activities will be on the north side of the Fermi site and all Fermi 3 construction activities will be on the south side of the site. The boundary separating Fermi 2 operations from Fermi 3 construction activities is roughly an east-west line extending across the site from the southern boundary of the Fermi 2 protected area. This constraint significantly reduces the amount of land available for building and construction because land north of the line will not be available for Fermi 3 construction.

11) Fermi 2 operations and the Fermi 3 construction site must have completely separate access roads, entrances and exits. Fermi 2 and Fermi 3 roads and activities must not cross each other. This is to avoid traffic impacting either site. This also relates to Constraint 7.

Constraints 2, 3, 4, 5, and 6 allow very little flexibility on where power block structures such as the reactor building can be located. The only location suitable is south of the existing Fermi 2 protected area on the opposite side of the imaginary east-west dividing line.

Constraints 7 and 8 require arranging the Fermi 3 site to ensure that there will be adequate space near the primary construction area to allow a free flow of personnel, materials and equipment. Fermi 3 requires a large construction workforce with up to 2900 construction workers at peak and 900 onsite workers when operational. Adequate staging and laydown area (temporary storage of construction materials) is needed to support the modular construction of nuclear power plants. Reactors such as the ESBWR proposed for Fermi 3, use standardized modules and certified designs to expedite the construction schedule. Nuclear power plant construction management consultants have advised Detroit Edison that a minimum of 100 acres of land should be available near the construction site for staging, laydown, and assembly of equipment and pre-assembled modules. A comparison of the amount of proposed land available for other United States nuclear

Appendix J

license applicants indicates that the Fermi 3 site, in the preferred site layout, is among the smallest sites in terms of acres used.

Constraint 9 requires a final review and approval of any proposed site layout arrangement by security subject matter experts with appropriate clearances to ensure that the layout is in compliance with all security plan requirements.

Efforts to minimize impacts in the alternatives development process included:

- Avoiding and minimizing impacts to all wetlands with priority given to avoiding impacts to the most valuable/functional wetlands;

- Where wetland impacts were unavoidable, the preference was for temporary wetland impacts over permanent wetland impacts, with the understanding that wetland mitigation implemented prior to, or concurrent with, the impact will still be required. A temporary impact means that the wetland will be restored to existing or better condition once the temporary land use for construction activities is completed; and,

- Placing the Fermi 3 power block in the largest contiguous upland area.

Efforts were made to avoid, to the extent practicable, adverse impacts associated with filling or modification of wetlands and new construction in wetlands wherever there is a practicable alternative. Impacts were only considered when there was no practicable alternative, and the proposed configuration for Fermi 3 includes all practicable measures to reduce impacts to wetlands and jurisdictional waters. Detroit Edison evaluated each of the onsite alternative layouts based on the approximate acreage, type, and value of wetlands that would be impacted. Alternatives that would minimize impacts to wetlands were preferred over alternatives that would result in greater impacts.

Wetland impacts of the Revision 0, Revision 1, and Revision 2 site layouts presented in the Fermi 3 Environmental Report, were evaluated using the updated Fermi site wetland delineation provided in this application (see Figure 2-2). Impacts to the open water areas H and U are treated as emergent wetland impacts. Therefore, the acres of impact presented here differ slightly from those presented in the Environmental Report.

Revision 0 Site Layout

Revision 0 is the site layout presented in the original Fermi 3 combined license application (COLA) submittal in September 2008. The Revision 0 layout was finalized in February 2008 using preliminary site wetlands information and was laid out along traditional concepts for large, long-term, construction sites.

Unchanged Site Layout Elements

The location of the Fermi 3 power block, which includes the reactor building, turbine building, control building, fuel building, radwaste building, diesel generators and other plant support systems, is fixed according to the requirements set out in Constraints 6 and 10. This location did not change in subsequent site-layout revisions.

Lake Erie will be used as the source for makeup water to the plant. The Fermi 3 makeup water intake will be adjacent to the intake for Fermi 2, i.e., located between the two existing groins that protrude into Lake Erie in the location of existing Fermi 1 structures. A barge slip for delivery of prefabricated modules, large components and building materials will be located between the two groins and adjacent to the south groin. These structures will be located in areas that have already been disturbed, in conformance with Constraint 1 and 10. The location of these structures did not change in subsequent revisions.

The Fermi 3 blowdown water outfall to Lake Erie will be offshore via an underwater discharge line in conformance to Constraints 1, 2 and 10. The configuration and discharge location of this line did not change in subsequent revisions. Four discharge locations were considered including two shoreline discharges (concrete, partially submerged, discharge structure along the shoreline) and an inland location. The inland location into the south lagoon was eliminated due to environmental considerations according to Constraint 1. The warm blowdown water could potentially disturb the local aquatic ecosystem and wetlands in the south lagoon. The two shoreline discharge locations considered on the south side of the site, per Constraint 2, were also eliminated due to environmental considerations per Constraint 1 and potential Fermi 2 operational impacts per Constraint 2. One consideration with both shoreline locations was the possibility of variable, near-shore currents sending the warm blowdown water back into the Fermi 2 and Fermi 3 makeup water intakes, which could impact plant heat loads and water chemistry. The other consideration with both shoreline locations was that warm blowdown water discharged during a seiche event, with winds from the east, could flow back into the south lagoon, potentially disturbing the local aquatic ecosystem and wetlands. Shoreline discharge locations would pose greater impacts than the proposed offshore discharge, which is considered environmentally preferable.

Site Layout Elements that Changed in Subsequent Site Layout Revisions

The normal power heat sink for Fermi 3 is a single concrete natural draft cooling tower. The cooling tower location changed from Revision 0 to Revision 1. Several criteria were utilized in identifying the initial cooling tower location, as follows:

- The cooling tower must be at least 800 feet away from safety-related structures in conformance with Constraint 9 (the cooling tower must be located, at minimum, a distance equal to its height from any safety-related structures such as the reactor building. This is to eliminate the potential for damage to these structures, if the tower collapsed), and;

- The cooling tower must be at least 1000 feet away from the switchyard to minimize icing and salt drift impacts also in conformance with Constraint 9.

Other considerations included the following: minimizing the length of the circulating water piping; minimizing the distance to Lake Erie, minimizing wetland impacts according to Constraint 1; minimizing Fermi 2 system impacts, and; minimizing temporary impacts to Fermi 2 and Fermi 3 site access during construction according to Constraints 2, 10 and 11. Four locations were considered. The location chosen was south of Fermi 3 in an area that was considered to be forested upland. The location selected conformed with the above-mentioned constraints and had the smallest impact to wetlands, the shortest circulating water pipe length, and had the smallest Fermi 2 system impacts.

In conformance with Constraints 10 and 11, several Fermi 2 operational facilities (warehouses, administration and engineering offices, maintenance shops) were relocated from the Fermi 3 construction site to the Fermi 2 side of the site. These facilities were to be relocated in an area that was considered to be forested upland. The location of these facilities changed from Revision 0 to Revision 1 to minimize wetland impacts, in conformance with Constraint 1, based on additional wetlands delineation information.

In conformance with Constraint 11, the Fermi 2 site to the north, and the Fermi 3 construction site to the south, must have completely separate access roads, entrances and exits. This is to prevent traffic from either site affecting the operation of Fermi 2 or Fermi 3. The Fermi 2 access road followed the west property line along Toll Road, then turned west through an area that was considered to be forested upland. The access road was altered from Revision 0 to Revision 1 to minimize wetland impacts, in conformance with Constraint 1, based on additional wetlands delineation information. The Fermi 2 access road was slightly altered in Revision 2 to further reduce wetland impacts.

The Fermi 3 temporary construction parking lot was proposed to be located on the north side of Fermi Drive, beneath the existing transmission corridors in accordance with the Fermi 2 and Fermi 3 separation requirements per Constraint 10. A large area is needed for construction parking to accommodate 2900 workers at the peak of construction. This area is also directly connected to the construction site and meets the requirements of Constraint 7. The utility of this

Appendix J

area for other construction activities was limited due to the existing high-voltage overhead lines. The location of construction parking and the utilization of this field changed from Revision 1 to Revision 2.

Revision 1 Site Layout

Based on completion of the Ducks Unlimited wetland study in July 2008, Detroit Edison recognized that the cooling tower location and the location of the Fermi 2 facilities moved from the Fermi 3 construction site, had greater wetland impacts than originally assessed and that these placements would have to be modified. Therefore, at the U.S. Nuclear Regulatory Commission (NRC) environmental audit in February 2009, Detroit Edison informed the NRC, Michigan Department of Environmental Quality (MDEQ), and the U.S. Army Corps of Engineers (USACE), that the Revision 0 site layout would be revised to further minimize wetland impacts.

Through planning and consultation with natural resource professionals, stakeholders and subject matter experts (nuclear security, materials management, construction planning, operations, maintenance, environmental and licensing), Detroit Edison developed a Revision 1 site layout that reduced wetland impacts to only those areas where a practicable alternative could not be identified that would still fulfill the overall project purpose. All available land onsite with no wetland impacts and low wetland impacts, that also conformed to the key constraints, was identified on a figure, for use in reconfiguring the Fermi 3 site layout. The stakeholder team then worked to eliminate or minimize wetland impacts by redesigning the site layout utilizing those identified low-impact and no-impact areas, with a focus on relocating Fermi 3 structures and activities with the greatest wetland impacts (e.g., cooling tower location, Fermi 2/Fermi 3 PAP, parking, office buildings, warehousing, and shops). The Revision 1 site layout was submitted to the NRC in December of 2009.

One of the key changes made to the Revision 1 site layout was moving the cooling tower from the forested wetland, south of Fermi Drive, to land just west of the Fermi 3 power block. This location has several advantages such as shorter circulating water lines, no temporary disturbance to construction site roadways, and no wetland impacts (per the 2008 wetlands delineation). One consideration of this location was that it was close to safety-related structures such as the reactor building. According to Constraint 9, the cooling tower was positioned a distance greater than its height from safety-related structures to prevent damage to these structures, if the tower were to collapse. The South Canal is impacted by the new cooling tower location and by the need to maintain a free flow of personnel, equipment and materials to the construction site, according to Constraint 7. The intersection of Fermi Drive, Quarry Lake Road and Doxy Road is considered a pinch point to the free flow of personnel, equipment and materials. Bridging of the South Canal allows for an unconstrained connection between the field to the west and the construction site. Due to the considerations explained above regarding Constraints 7 and 9, the impact to the South Canal is unavoidable.

A disadvantage to locating the cooling tower adjacent to the Fermi 3 power block is the loss of a large expanse of land adjacent to the primary construction site needed for laydown, staging, fabrication and assembly of modular components, according to Constraint 8. This loss can be partially, but not completely, compensated by managing the construction sequence. To address this constraint, the area known as the "pork chop" located south of Fermi Drive and west of Quarry Lakes Road, was utilized in the Revision 1 site layout, in conformance with Constraints 7, 8, and 10. The "pork chop" provides approximately 30 acres of prime construction land that includes 11.80 acres of forested wetland near the construction site. Natural resource inventories suggested the forested wetland in this area was of lower value ecologically than the other large forested systems onsite. The wetland is connected hydrologically with culverts but fragmented from other wetland areas and Lake Erie due to multiple roadways completely surrounding the site. It also had a larger component of dead/dying ash trees and invasive species and was subject to ongoing disturbance.

The "pork chop" is an important feature of the Revision 1 site layout due to its proximity to the construction site; location adjacent to Fermi Drive and rail access; and, the absence of overhead

Revision 1   Page 7 of 15   August 2011 transmission lines that can present a safety hazard and barrier to movement and assembly of equipment, materials and modules. Construction warehouses, staging, assembly areas, and maintenance shops were planned for this location. Utilization of this area greatly facilitates the free flow of personnel, equipment and materials, further relieving the pinch-point concern at the Fermi Drive and Quarry Lakes Road intersection. Traffic through this area includes workers and materials coming from Dixie Highway, laydown and staging areas, the rail spur, and the barge slip.

The other key change to the Revision 1 site layout was removing the Fermi 2 operational structures (permanent parking lot, warehouses, an administration building and maintenance shops) from the forested wetland west of the Fermi 2 protected area. These structures were relocated in the Revision 1 site layout as follows:

- An administrative support campus outside the owner controlled area, associated with the Nuclear Operations Center/Nuclear Training Center (NTC), was created to move the Fermi 2/Fermi 3 Administration Building and the Fermi 3 Training Simulator out of forested Wetland I, in conformance with Constraint 1. Conformance to Constraints 4, 10 and 11 was evaluated for this location due to Fermi 2 operational support facilities being moved to the southern, Fermi 3 side of the site. Several considerations mitigate these constraint conformance issues, as follows: a bridge or tunnel will be utilized to cross Fermi Drive without affecting the construction site; personnel utilizing the training facility and administrative offices are generally at that location the entire day and would not need to cross to the Fermi 2 side of the site; and, increased use of technology such as video conferencing will minimize cross over. In addition, this arrangement reduces the need for additional operational parking at the PAP due to reduced personnel inside the protected area, which reduces the parking-structure foot print, thus minimizing environmental impacts in this area in conformance with Constraint 1.

- The flat operational parking was moved out of forested Wetland I and replaced by two multiple-level parking structures to minimize land use and wetland impacts, and to improve the overall site parking situation in conformance with Constraint 1. One parking structure is proposed near the NTC for permanent training and administration parking to support the new administrative campus. The other structure is located near the new PAP on the west side of the protected area boundary for protected area parking. A small wetland impact associated with a portion of this parking structure remains. This impact could not be avoided due to the proximity of existing and proposed structures in this area, along with nuclear security distance requirements in conformance to Constraint 9. The two parking garages will be sized to accommodate Fermi 2 and Fermi 3 operational parking.

- The combined Fermi 2/Fermi 3 warehouse was moved out of forested Wetland I in conformance with Constraint 1 and moved east to straddle the protected area boundary near the vehicle inspection building (VIB) and PAP. This location minimizes impacts, however some wetland impacts were unavoidable due to necessary sizing of the Fermi 2/Fermi 3 warehouse and the need for an access road along the west side of the structure. This arrangement will improve operational efficiency of the Fermi 2 and Fermi 3 sites. Other areas north and west of the protected area were considered, however, key stakeholder feedback, primarily from materials management and nuclear security, insisted on this location for secure protected area operations in conformance with Constraints 2, 3, 6 and 9. Two other smaller warehouses (32 and 34) were also moved out of forested Wetland I, to a location along the access road with no associated wetland impact.

- The Fermi 2 operational access road was moved to minimize environmental impacts in conformance with Constraint 1. The access road no longer cuts through forested Wetland I. The access road now follows the existing Toll Road, then transitions to existing site roads, which route around Wetland I to access the site. Wetland impacts were minimized, however some impacts were unavoidable, in conformance with Constraints 6, 10 and 11. The unavoidable impacts were associated with a new Fermi 2 operational security gate,

Appendix J

necessary road improvements and rerouting of the existing road along the west side of the new Fermi 2/Fermi 3 warehouse.

Other modifications reflected in the Revision 1 site layout include the following:

- The Fermi 2/Fermi 3 meteorological tower was relocated because the new Fermi 3 cooling tower location will interfere with the current meteorological tower location. The new meteorological tower is relocated in an area near the southeast corner of the site. This location was selected because there were no known wetland impacts in conformance with Constraint 1 and because it met NRC regulatory guidance for meteorological tower placement.

- Construction staging and laydown was added on the south site border in a low-wetland impact area, on the east side of Quarry Lakes Road and around Fox Road, in conformance with Constraints 8 and 10. Unavoidable, temporary impacts are incurred to several small, fragmented, low-value emergent and scrub shrub wetlands (Wetlands AA, JJ, II). Nuclear construction subject matter experts engaged by Detroit Edison indicated that more land was needed for construction activities (staging, laydown, temporary spoils storage, and component assembly) than was originally allocated in the Revision 0 site layout.

- The Fermi 3 switchyard was moved to the agricultural field at the far west side of the property, adjacent to the south side of Fermi Drive. In Revision 0, the Fermi 3 switchyard was adjacent to the Fermi 2 switchyard in the protected area. Further analysis of the Fermi 3 interconnection determined the available space adjacent to the Fermi 2 switchyard was not sufficient for the new Fermi 3 switchyard. In addition, in accordance with Constraint 2, the original location was an impediment to movement and a potential impact to Fermi 2 operations. The new location also places the switchyard outside the owner-controlled area to facilitate access by ITC*Transmission* (owner and operator of the switchyard).

Revision 2 Site Layout

After the Revision 1 site layout was finalized, terrestrial and aquatic studies continued on the site. The results indicated a greater diversity in the vegetative communities within the "pork chop," than was originally understood. Subsequently, in a meeting to discuss Fermi 3 wetland permitting in July 2010, the MDEQ and USACE indicated that the wetland impacts associated with the "pork chop," contained in the Revision 1 site layout, were problematic. In response to this feedback and in conformance with Constraint 1, Revision 2 of the site layout was developed to address the wetland impact to the "pork chop" area.

Construction activities were moved out of the "pork chop" (Wetlands BB, EE, and FF) and the contiguous forested upland associated with that parcel, in accordance with Constraint 1. Site elements were rearranged to eliminate the "pork chop" impact, in conformance with Constraints 1, 7, 8 and 10. Most of the construction activities planned for the "pork chop," were moved to the north side of Fermi Drive. Some of the construction activities were also moved into areas designated for construction laydown located around the Quarry Lakes. Construction parking originally planned for the field north of Fermi Drive, was moved into the farmer's field located along the western property line. The use of the field on the north side of Fermi drive was limited in the previous site layout because of existing overhead transmission lines, so in Revision 2, the 345 kV lines are rerouted.

The resulting changes are summarized as follows:

- The 345 kV transmission lines that serve Fermi 2 and the proposed Fermi 3 were rerouted to open up the field on the north side of Fermi Drive for all necessary construction activities to satisfy Constraints 7, 8 and 10. The transmission is rerouted due west through emergent Wetland C, then south along Toll Road, to the Fermi 3 switchyard, which was moved into the field at the corner of Toll Road and Fermi Drive. This change eliminates impacts to a large parcel of rare and imperiled wetland (the "pork chop") and incurs unavoidable impacts to approximately 2 acres of forested wetland (the impacts will change the edge of Wetland F below the transmission lines from a forested wetland to a emergent wetland) and small, unavoidable, permanent and temporary impacts to an emergent Wetland C.

- Land surrounding the Quarry Lakes, designated as laydown, was added for various construction activities in conformance with Constraints 7, 8 and 10, to replace loss of laydown and staging areas from the "pork chop" area and from moving construction parking into the farmer's field. Some temporary, unavoidable impacts are incurred to small, fragmented, low-value forested and emergent wetlands in these areas (Wetlands W and Y).

- The Fermi 3 switchyard was moved from the south side to the north side of Fermi Drive to facilitate the transmission corridor rerouting in conformance with Constraints 1, 7 and 8. Construction parking, previously located in the field north of Fermi Drive, is moved into the farmer's field.

- The Fermi 2 access road was realigned to further minimize impacts to forested Wetland I in conformance with Constraint 1. The new alignment will follow Toll Road further north, just past Langton Road, prior to transferring onto the Fermi site access road.

- The meteorological tower was moved southeast of the Revision 1 location to eliminate any potential wetland impacts. When the Revision 1 location was identified, the understanding was that cutting trees in a wetland did not require a wetland permit. At the July 2010 meeting with the MDEQ and USACE, the staff clarified that cutting trees from forested wetland areas in association with the meteorological tower would require a permit for the conversion of wetland type. In conformance with Constraint 1, the Revision 2 site layout identified a location that was consistent with the recommendations of the meteorological tower siting study and did not require tree cutting in wetland areas.

- In Revision 2, construction boundaries were refined to eliminate unintended impacts in the Revision 1 site layout associated with construction along Quarry Lake Road and the Dredged Spoils Disposal Basin.

- Operations and maintenance dredging authorized under existing Fermi 2 permits was eliminated as an impact attributed to Fermi 3 construction (reduction of 7.32 acres of open water impacts). The incremental change in the extent of dredging within Lake Erie required to support Fermi 3 construction was included.

Preferred Site Layout

Refinements to the Revision 2 site layout were made during the development of the joint permit application. Detroit Edison modified the alignment of the new operations access road to avoid potential wetland impacts in the area west of the existing Toll Road. This change resulted in a small increase in the forested and emergent wetland impacts on the Fermi property side of the access road. The shift in the access road alignment altered the path of the onsite transmission, resulting in an increase of 1 acre (from 1.53 acres to 2.53 acres) in the forested wetland that would be cleared within the transmission corridor. The proposed roadway, security gate, and box culvert design were modified to minimize the encroachment into the wetland areas as much as practicable. Overall the wetland impacts associated with the road increased by 0.53 acre. The wetlands west of the existing Toll Road have not been formally delineated. Based on federal wetland mapping and field observations, Detroit Edison believes equal or greater wetland impacts would have resulted from the previous access road alignment.

Summary of Project Alternatives and LEDPA Analysis

Table 4-3 compares potential impacts to wetlands on the Fermi site of the four alternative site layouts discussed above. Wetland impacts were further characterized by Michigan Natural Communities to illustrate impacts to higher valued wetlands.

Appendix J

Detroit Edison minimized potential project impacts to waters of the United States, including wetlands. The site layout for the Fermi 3 project was based on an iterative approach to determine a layout that would most practicably avoid and minimize impacts to USACE jurisdictional waters and wetlands. Areas of the Fermi site that represented no, or minimal, impacts to wetland functions and values were identified. Stakeholders were engaged to identify constraints on the site layout, including integration of Fermi 3 with the ongoing operations of Fermi 2. Those constraints were used to identify locations for the proposed Fermi 3 and associated construction. Efforts were made to avoid, to the extent possible, impacts associated with the destruction or modification of wetlands and streams and new construction in wetlands and streams wherever there was a practicable alternative.

The Fermi 3 power block was located in the largest contiguous upland area consistent with Constraints 1, 2, 3, 4, 5, 6, 7, 9 and 10. The cooling tower was also located in this upland area at a distance from the power block that satisfies nuclear safety considerations, per Constraint 9. The minimum separation distance precludes siting the cooling tower entirely within the available upland adjacent to the Fermi 3 power block area.

A combined Fermi 2/Fermi 3 warehouse, parking, VIB, and PAP located on the west side of the protected area boundary, offers significant efficiency advantages over the operational life of the plants. A multi-level parking structure connected to the PAP addresses the need for parking for an additional 900 staff when Fermi 3 is operational while minimizing impact to the adjacent wetlands. The location of these facilities supports the integration of the Fermi 2 and Fermi 3 protected areas when construction is completed and satisfies other nuclear security considerations per Constraints 2, 3, 6, 9 and 10.

Construction of the Fermi 3 intake structure, discharge pipe, and barge slip within the existing Fermi 2 intake embayment reduces the cumulative area of lake bottom that will be disturbed per Constraint 1. The discharge pipe and fish return pipe are the only Fermi 3 components that will require dredging beyond the operations and maintenance dredging currently authorized for Fermi 2 under MDEQ and USACE permits.

Adequate laydown area is needed to support the modular construction that is a key component of modern nuclear power plants, as described in Constraint 8. Reactors such as the ESBWR proposed for Fermi 3 use standardized modules to expedite the construction schedule. With the relocation of the 345kV transmission, the field to the west, and immediately adjacent to the power block, along the north side of Fermi Drive, possesses the attributes necessary for key construction activities consistent with Constraints 7 and 8. Use of this area includes some unavoidable impacts to wetland areas that will be restored following completion of construction of Fermi 3.

The design iterations reduced the potential wetland impacts from over 150 acres to approximately 40 acres. Overall impacts to wetlands were reduced in the Preferred Alternative. Open water impacts were also reduced in the Preferred Alternative. The Preferred Alternative also reduces the total impact to those Michigan Natural Communities that are considered rare and imperiled. These include Great Lakes marsh and southern swamp (southern hardwood swamp). All the permanent and temporary wetland impacts in the preferred site layout were unavoidable given the ten constraints previously outlined. The preferred alternative presents significantly less impact to the high functioning, high value wetland communities at the Fermi site. Based on the results of the alternative site layout analysis, the Preferred Alternative was selected as the proposed site layout that best addresses avoidance and minimization of wetland impacts.

Table 4-1. Candidate Site Practicability Review (Sheet 1 of 2)

	Site A – Petersburg	Site C – South Britton	Site F – Greenwood	Site M – Fermi	Site N – Belle River	Site W1 – Port Austin	Site W2 – Caseville	Site W3 – Bay Port
Land Acquisition	Impracticable. 32 private owners, few houses.	Acceptable. 14 private owners, 15-25 houses/ facilities. May need to acquire additional land for EAB	Acceptable. Detroit Edison. Would need to acquire additional land for EAB	Acceptable. Detroit Edison. Fermi 3 EAB entirely within existing Fermi property and security zone	Acceptable. 81% Detroit Edison / 19 % Michigan Public Power Authority	Impracticable. 85 private owners. Many houses/ facilities	Impracticable. 90 private owners. Many houses/ facilities	Impracticable. 120 private owners. Many houses/ facilities. May need to acquire additional land for EAB
Transmission Lines	Acceptable. 345-kV lines with available capacity 1.2 miles north of site	Acceptable. 345-kV line with available capacity 1 mile north of site	Marginal. 345-kV line onsite but congested	Acceptable. 345-kV line with available capacity onsite	Marginal. 345-kV line onsite but congested	Impracticable. Nearest 345-kV line is approximately 48 miles from the site	Impracticable. Nearest 345-kV line is approximately 41 miles from the site	Impracticable. Nearest 345-kV line is approximately 35 miles from the site
Water Supply	Impracticable. 15.4 miles inland from Lake Erie	Impracticable. 24.4 miles inland from Lake Erie	Acceptable. 11 miles inland from Lake Huron	Acceptable. On the shore of Lake Erie	Acceptable. 2 miles west of St. Clair River	Acceptable. 1.4 miles inland from Lake Huron	Acceptable. 2.8 miles inland from Lake Huron	Acceptable. 1.4 mile inland from Saginaw Bay

Revision 1

Page 12 of 15

August 2011

January 2013

J-15

NUREG-2105

Table 4-1. Candidate Site Practicability Review (Sheet 2 of 2)

	Site A – Petersburg	Site C – South Britton	Site F – Greenwood	Site M – Fermi	Site N – Belle River	Site W1 – Port Austin	Site W2 – Caseville	Site W3 – Bay Port
Hazardous Land Uses	Impracticable Petroleum product pipeline 2 miles south. Two natural gas pipelines traversing the site from southwest to northeast within ½ mile of plant	Impracticable Two natural gas pipelines traversing the site from southwest to northeast. Would require relocation of a 30-inch line to avoid conflicts with the plant	Marginal Oil-fired peaking unit and three gas turbines onsite	Acceptable Two limestone quarries 3 miles northeast.	Impracticable Multiple large natural gas transmission lines, gas storage field and compressor station within 2 miles. Bulk petroleum facility 3 miles north of the site	Acceptable No hazardous land use sites within 5 miles.	Acceptable No hazardous land use sites within 5 miles.	Acceptable Limestone quarry and anhydrous ammonia facility within 3 miles of the site.
Railroad Access	Acceptable Indiana & Ohio Railroad 1.5 miles west of the site.	Acceptable Norfolk Southern Railway 1.9 miles east of the site.	Acceptable PVTX Railway spur on site.	Acceptable Canada National Railway spur on site.	Acceptable CSX Transportation spur on site.	Acceptable Huron & Eastern Railway 1.4 miles southeast of the site.	Marginal Huron & Eastern Railway 6.7 miles south of the site.	Acceptable Huron & Eastern Railway 5.4 miles south of the site.
Overall Conclusion	Impracticable	Impracticable	Acceptable	Acceptable	Impracticable	Impracticable	Impracticable	Impracticable

Table 4-2. Comparison of Wetland/Water Impacts from Alternative Sites

Onsite Wetlands/Waters	Proposed Site Fermi		Alternative Site Greenwood	
Delineated Property Acreage	1106		1729	
Wetlands Acreage	513		386	
Open Water Acreage	40		NA	
Streams Linear Feet (LF)	0		30,303	
Wetlands Affected Acreage	40		39	
Streams Affected LF	0		401	
Open Water (Lake Erie) Affected Acreage	0.08		NA	
Open Water (inland) Affected Acreage	NA		NA	
Offsite Wetlands/Waters	Wetlands (acreage)	Streams (LF)	Wetlands (acreage)	Streams (LF)
Makeup Water Intake (acreage)[a]	-	-	NA	NA
Water Pipeline ROW	-	-	3.1	4378
Transmission Line ROW	121	7304	257	29,648
Blowdown Pipeline ROW	-	-	0	273
Total Wetlands/Waters Affected				
Wetlands Affected Acreage	161		300	
Streams Affected LF	7304		34,701	
Open Water (Lake Erie) Affected Acreage	0.08		NA	
Open Water (inland) Affected Acreage	NA		NA	

[a] Impacts within Lake Huron for the construction of an intake structure for the Greenwood site alternative were not evaluated.

Appendix J

Table 4-3. Comparison of Impacts for Alternative Site Layouts

Type	Revision 0	Revision 1	Revision 2	Preferred Alternative
Wetland Impacts (acres) by Type				
PEM wetland[a]	54.84	18.79	26.08	26.40
PFO wetland	96.66	18.97	6.84	8.03
PSS wetland	7.00	4.10	5.28	5.28
Total wetlands	158.49	41.86	38.19	39.71
Open water	8.87	7.40	0.08	0.08
Wetland Impacts (acres) by Michigan Natural Community[b]				
Rare and imperiled: Great Lakes marsh	47.53	10.38	12.86	13.19
Rare and imperiled: southern hardwood swamp	92.19	14.08	1.95	3.15
Southern shrub carr	7.00	3.92	3.91	3.91
PEM wetland – coastal	0	0.80	0.80	0.80
PEM wetland[a]	7.31	7.61	12.42	12.42
PFO wetland	4.47	4.89	4.89	4.89
PSS wetland	0	0.18	1.37	1.37
Open water	8.87	7.40	0.08	0.08

[a] Includes 1.88 acres of nonjurisdictional PEM wetland impacts.

[b] Chapter 324, Section 303.01(t) of the Michigan Natural Resources and Environmental Protection Act lists Michigan Natural Communities that are considered rare and imperiled. These include Great Lakes marsh and southern swamp (southern hardwood swamp). Any wetland considered "other" that is connected hydrologically to Lake Erie or is within 1000 feet of the ordinary high water mark (elevation 571.6 feet IGLD 1955) is considered coastal.

J.3 References

33 CFR Part 320. Code of Federal Regulations, Title 33, *Navigation and Navigable Waters,* Part 320, "General Regulatory Policies."

40 CFR Part 230. Code of Federal Regulations, Title 40, *Protection of Environment,* Part 230, "Section 404(b)(1) Guidelines for Specification of Disposal Sites for Dredged or Fill Material."

Detroit Edison Company (Detroit Edison). 2011. *Detroit Edison Fermi 3 Project, U.S. Army Corps of Engineers and Michigan Department of Environmental Quality, Joint Permit Application.* Revision 1, Detroit Michigan. August. Accession No. ML112700388.

U.S. Army Corps of Engineers (USACE). 2011. "Public Notice: Proposed Structures and Dredge and Fill Activities Associated with the Proposed Enrico Fermi Unit 3 Nuclear Power Plant in Lake Erie and/or Adjacent Wetlands at Frenchtown Charter Township, Monroe County, Michigan." Accession No. ML12180A374.

Appendix K

Detroit Edison's Proposed Compensatory Mitigation Plan for Aquatic Resources

Appendix K

Detroit Edison's Proposed Compensatory Mitigation Plan for Aquatic Resources

This appendix presents Detroit Edison Company's (Detroit Edison's) proposed plan to compensate for its proposed unavoidable adverse impacts to aquatic resources associated with the building of Enrico Fermi Unit 3 (Fermi 3), as presented in its onsite alternatives analysis (Appendix J).

Based on guidance provided by the U.S. Army Corps of Engineers (USACE) during pre-application coordination regarding Clean Water Act Section 404(b)(1) Guidelines compliance, Detroit Edison conducted an onsite alternatives analysis (Detroit Edison 2011), contained in Appendix J, and identified its proposed least environmentally damaging practicable alternative (LEDPA) to avoid and minimize impacts on waters of the United States. Since Detroit Edison's proposed LEDPA would result in unavoidable adverse impacts to aquatic resources, Detroit Edison initially developed a conceptual-level mitigation strategy (Detroit Edison 2011) as a starting point to address the required compensatory mitigation for the unavoidable losses attributable to its LEDPA. Detroit Edison's proposed LEDPA and conceptual mitigation strategy were made available for public review and comment in Appendices J and K, respectively, of the Draft EIS. The USACE LRE-2008-00443-1-S11 public notice ending January 23, 2012 (USACE 2011), provided additional opportunity for public comment on both the proposed LEDPA and the conceptual mitigation strategy.

As discussed in Appendix J, USACE is actively reviewing and coordinating with Detroit Edison regarding its proposed LEDPA. This is part of the ongoing USACE process to identify and verify the USACE LEDPA and determine compliance with other restrictions of the Guidelines and public interest review. Subsequent to the Draft EIS and USACE public notice, and based on USACE comments and coordination regarding its conceptual mitigation strategy, Detroit Edison refined and detailed its mitigation strategy and produced the proposed mitigation plan that is now contained in this appendix. USACE is actively evaluating this proposed plan in conjunction with the proposed LEDPA. The final mitigation plan must be approved by the District Engineer prior to USACE issuance of a permit for the proposed work related to the Fermi 3 project. A USACE permit, if issued, would include special conditions that would state the compensatory mitigation requirements including the amount and type of compensatory mitigation; identify the responsible party for providing the compensatory mitigation; incorporate, by reference, the final mitigation plan approved by the USACE District Engineer; and unless provided in the approved final mitigation plan, describe, for the compensatory mitigation project, the required financial assurances and long-term management provisions, plan objectives, required monitoring, and performance standards, which include Detroit Edison's confirmation that the mitigation meets the Federal wetlands criteria as discussed in Section 1.1.3 of this EIS.

Fermi 3
U.S. Army Corps of Engineers Mitigation Strategy and Final Design

MDEQ/USACE Joint Permit Application

PREPARED BY:
CONSERVATION CONNECTS
TETRA TECH

July 2012

Fermi 3 USACE Mitigation Strategy and Final Design

TABLE OF CONTENTS

i

ii

Appendix K

iv

Appendix K

1.0 INTRODUCTION

Detroit Edison has developed the following mitigation strategy to compensate for proposed impacts to aquatic resources associated with construction of Fermi 3 (Proposed Development) at the Enrico Fermi Atomic Power Plant (Fermi site). The Proposed Development site is located on the western shore of Lake Erie at Newport, Monroe County, Michigan on a 1,260-acre parcel owned and managed by Detroit Edison (Figure 1).

A full description of the Proposed Development was presented in the associated Joint Permit Application [Michigan Department of Environmental Quality (MDEQ) File Number 10-58-0011-P, U.S. Army Corps of Engineers (USACE) File Number LRE-2008-00443-1-S11]. Proposed impacts include 35.55 acres of mixed wetland types within the coastal zone of Western Lake Erie and the northern portion of the Ottawa-Stony Watershed, USGS Cataloging Unit and Hydrologic Unit Code (HUC): 04100001. Wetland types are classified broadly according to the U.S. Fish and Wildlife Service (USFWS) Cowardin classification and more specifically according to the Michigan Natural Community classification. Potential impacts include approximately 10.90 acres of palustrine emergent marsh (PEM; Great Lakes marsh), 3.15 acres of palustrine forested wetland (PFO; southern hardwood swamp), 3.91 acres of palustrine scrub shrub (PSS; southern shrub carr), 0.80 acres of PEM (coastal emergent wetland), 10.53 acres of PEM (other emergent wetland), 4.89 acres of PFO (other forested wetland) and 1.37 acres of PSS (other scrub shrub wetland).

To compensate for the wetland impacts, Detroit Edison proposes to restore wetlands offsite in the coastal zone of Western Lake Erie. This mitigation strategy is based on data collected onsite, existing databases, the attributes of potentially impacted wetlands, watershed priorities, feedback from natural resource professionals and ongoing communication with the regulatory and conservation community.

2.0 MITIGATION GOALS AND OBJECTIVES

The principal goal of this mitigation strategy is to restore and protect wetland functions and services of equal or greater value than those impacted by construction of the Proposed Development (Figure 2). This goal will be achieved through offsite wetland mitigation activities within the coastal zone of Western Lake Erie. The specific objectives listed below were developed based on an in-depth evaluation of the natural resources at the impact site and the mitigation site, and the condition and conservation needs of the surrounding watershed (see Section 3.1). A watershed analysis allowed for integration of watershed attributes including history, current condition, land use trends, stressors, conservation priorities and other conservation efforts in the Ottawa-Stony watershed and the coastal zone of Western Lake Erie in Monroe County, Michigan (Section 3.1.9). Site level and landscape level perspectives were combined with feedback from regulatory and conservation agency staff to develop an integrated compensation strategy, consistent with guidance from the USACE contained in 33 CFR Part 332 – Compensatory Mitigation for Losses of Aquatic Resources, the Environmental Protection Agency guidance contained in 40 CFR Part

1

230 – Section 404(b)(1) Guidelines for Specification of Disposal Sites for Dredged or Fill Material, and the MDEQ Technical Guidance for Wetland Mitigation (Reference 1).

2.1 Mitigation Overview

Over 500 acres of wetlands are present at the Fermi site. Wetlands potentially impacted by the Proposed Development have been avoided and minimized to the maximum extent practicable. Aquatic resources on the Fermi Site were identified, evaluated and considered throughout the design process. The first consideration was to determine if wetland impacts could be avoided entirely. The second consideration was to minimize potential impacts in terms of both quantity and quality to the maximum extent possible. The third consideration was to develop a mitigation strategy that would compensate for all unavoidable impacts. Design iterations reduced potential wetland impacts from over 150 acres to approximately 35.55 acres of regulated[1] wetlands requiring mitigation (21.4 acres of which will be restored post-construction). In addition to reducing total acreage of impacts, wetland location and quality were taken into consideration as discussed below and in Section 3.1.

To compensate for the loss of wetlands at the Proposed Development site, Detroit Edison will restore wetlands of similar ecological type within the same coastal zone. For the purposes of this document, restoration implies re-establishing conditions under which the natural functions of a pre-existing wetland can recover. To achieve the mitigation goal stated above Detroit Edison will restore wetlands offsite in the coastal zone of Western Lake Erie (Figure 3).

This comprehensive mitigation strategy is unique in that it proposes mitigation that will ultimately restore significant coastal wetland resources with direct connection to lake hydrology along Lake Erie. Detroit Edison proposes to implement these conservation measures to satisfy the site-specific compensation requirements for impacts to wetlands and address critical watershed needs and priorities as described below in Section 3.1.9. Mitigation activities will commence prior to or concurrent with wetland impacts at the Fermi site to reduce temporal losses of aquatic functions.

Under Part 303, Wetlands Protection, of the Natural Resources and Environmental Protection Act, 1994 PA 451, as amended, MDEQ may require compensatory wetland mitigation to replace unavoidably lost wetland resources with created or restored wetlands, with the goal of replacing as fully as possible the functions and public benefits of the impacted wetlands. A functional assessment was conducted to evaluate individual wetlands potentially impacted by the Proposed Development and to define appropriate compensation. A wetland mitigation and monitoring plan detailing the proposed mitigation activities has been submitted to MDEQ in accordance with Permit 10-58-0011-P. The proposed wetland restoration described herein satisfies the MDEQ requirements for wetland mitigation as set forth in the permit.

---

[1] Regulated wetland acreage includes those wetlands regulated by USACE and/or MDEQ.

2

2.2 Functional Replacement and Functional Lift

Restoration activities emphasize heterogeneity in microtopography, vegetation and hydrology to maximize diversity and ecological resilience of wetland habitat. Wetland mitigation has been designed to specifically replace the functions and values provided by wetlands with proposed impacts at the Fermi site. These functions and values include varying degrees of flood flow attenuation and storage, sediment, nutrient and toxicant retention, and fish and wildlife habitat. Section 3.1.8 details the wetland conditions, functions and values of impacted wetlands. Wetland mitigation has also been designed to significantly increase aquatic functions at the mitigation site over the level currently provided by existing wetlands. Existing wetlands are actively farmed or exhibit varying degrees of disturbance to hydrology, invasive species and disturbance from adjacent agricultural activities. Section 3.2 and Reference 38 describes the existing conditions of the mitigation site. The final mitigation design targets functions and values of high priority to the surrounding watershed including food chain support, breeding and migration habitat for migratory birds, breeding and over-wintering habitat for amphibians, increased nutrient cycling, increased connectivity of habitat types, and water quality improvements for surface outflow to Lake Erie.

The Evaluation of Planned Wetlands (EPW) method (Reference 2) was used to quantify the expected functional replacement of wetlands and the functional lift expected at the mitigation site. The EPW method focused on two comparisons. The first comparison describes and estimates how wetland functions provided by the planned wetland restoration (planned wetland) at the mitigation site compares to the lost functions of wetlands at the Fermi 3 site (impact wetlands). The second comparison quantifies the projected functional lift at the mitigation site by comparing projected wetland functions provided by the planned wetland to existing wetland functions at the mitigation site (Monroe Wetlands, Reference 37).

The EPW method was selected for several reasons. First, in the absence of a quantitative or scoring wetland assessment method for the Detroit District, the EPW provides a rapid assessment method based on a generic ecological model with the intention that it be applied to wetlands in the United States regardless of location. Second, the EPW method was developed specifically to evaluate projected functional values for planned wetlands. This evaluation provides guidance on final design and determines the degree of likelihood that mitigation requirements will be met. Finally, the EPW has been used by USACE and other state and federal agencies to evaluate wetland restoration and mitigation projects in New York, Maryland, Delaware and Virginia, many of which were as large and complex as Fermi 3.

Wetland functions and conditions of impact wetlands and current conditions of the mitigation site as assessed in the field compared with the targeted functions of the planned mitigation wetland demonstrate that the planned wetland is designed to specifically replace lost functions at the impact area and significantly improve on functions currently provided by wetlands at the mitigation site. The EPW method utilized previous assessment data and resulted in functional capacity calculations and comparisons that provide a clear, numerical description of how the mitigation action compensates for unavoidable impacts

3

to wetlands at the Fermi site and provides significantly increased benefits at the mitigation site. For each function evaluated (sediment stabilization, water quality, wildlife habitat, fish habitat, unique/heritage), the planned wetland matched or exceeded the functional capacity index of the impact wetlands and the existing conditions of the mitigation site. Weighted by area, the planned wetland is shown to significantly increase functional capacity over the impacted wetlands and over the functional capacity of the wetlands that currently exist at the mitigation site. The functional capacity of the planned wetland also exceeded the primary planned wetland goal which was to replace lost wetland functions of impact wetlands at an average replacement ratio of 3:1. The evaluation assumes the functional capacity of the impacted wetland is permanently lost; however, approximately 60% of the wetland impacts are temporary and the functions and values associated with those wetlands would be restored post-construction.

Based on field assessments and functional analysis, the mitigation plan is expected to exceed replacement goals for all wetland impacts and provide significant functional lift at the mitigation site. It is recognized that there is typically a time lag between loss of wetland functions due to wetland impacts and the gain of wetland functions at the mitigation site. As stated above, mitigation activities will commence prior to or concurrent with impacts to reduce temporal loss. The additional functional capacity projected for the planned wetland over and above impact wetlands, existing mitigation site wetlands and stated wetland goals will provide further compensation for temporal loss associated with both temporary and permanent impacts at the Fermi site.

2.3 Mitigation Acreages

A summary of wetland impacts and attributes is provided in Table 1. A more detailed description of the impacted wetlands is provided in Section 12 of the associated Joint Permit Application.

Wetland mitigation proposed here will replace wetland functions and values impacted on the Fermi site by restoring approximately 130 acres of wetlands of similar type offsite in the same watershed (coastal zone). Restoration will include approximately 97 acres of Great Lakes marsh (which includes 70 acres of emergent and 27 acres of open water), 22 acres of PFO (southern hardwood swamp), and 11 acres of PSS wetland. Table 2 provides the types and acreages of wetlands impacted and the proposed acreage of mitigation. Figure 4 shows the derivation of the mitigation acreages. In addition, the onsite restoration of 21.4 acres of the impacted wetlands post-construction will provide added ecological value and benefits above the required compensatory mitigation.

In summary, Detroit Edison recognizes the value of coastal wetland habitat along Lake Erie. Avoidance and minimization strategies were employed to minimize impacts to wetlands of high ecological value. Unavoidable impacts were restricted to low quality wetlands and wetland areas to the greatest extent possible. As described above, each acre of wetland impacted will be compensated for by the restoration of approximately 3 acres of high quality, intact wetland with a significantly greater projected functional capacity than impact wetlands and existing aquatic resources at the mitigation site. Additional compensation will be realized by post-construction restoration of approximately 60% of the impacted

4

wetlands onsite. This mitigation strategy satisfies regulatory mitigation requirements with proposed compensation at an appropriate level to achieve replacement of lost functions and values including temporal loss of aquatic resource functions. This mitigation strategy will also support Detroit Edison's corporate environmental stewardship initiatives through continued collaboration and partnership with USFWS and other conservation entities.

3.0 BASELINE INFORMATION

3.1 Impact Area

3.1.1 Location and Ownership

The Proposed Development is at the Fermi site, Latitude: 41.961 and Longitude: -83.261 on the western shore of Lake Erie at Newport, Monroe County, Michigan on a 1,260-acre parcel owned and managed by Detroit Edison (Figure 1). The impact site is within the coastal zone of Western Lake Erie and the northern portion of the Ottawa-Stony Watershed.

3.1.2 Land Use

Land use on the Fermi site is split mainly into developed areas and swamp or wetland areas. Most of the forested areas on the site are subject to flooding, and, therefore, are considered woody wetlands. The majority of the Fermi site that is not developed is included as part of the Detroit River International Wildlife Refuge (DRIWR), known as the Lagoona Beach Unit. The DRIWR encompasses a 656-acre portion of the Fermi site.

The 1260 acre Fermi site is composed of approximately 16.8% developed areas and 5.1% cropland. Terrestrial habitats account for 61% of the property. The remaining 17% are water bodies, e.g., Quarry Lakes and the main body of Lake Erie that lies east and north of the site. Figure 5 illustrates the extent and location of the habitats identified and the developed areas on the Fermi site. A summary of the acres of each habitat type on the site is provided below (Reference 7).

5

Habitat	Acres	Percent of Site
Coastal Emergent Wetland Open Water	35	2.8
Coastal Emergent Wetland Vegetated	238	18.9
Grassland: Right-of-Way	29	2.3
Grassland: Idle/Old Field/Planted	75	6.0
Grassland: Row Crop	64	5.1
Shrubland	113	9.0
Thicket	23	1.8
Forest: Coastal Shoreline	47	3.7
Forest: Lowland Hardwood	92	7.3
Forest: Woodlot	117	9.3
Developed Areas	212	16.8
Lakes, Ponds, Rivers	44	3.5
Lake Erie (main body)	171	13.6
Totals	1,260	100.0

3.1.3 Topography

Topography in the vicinity is fairly flat, with some lower elevation wetland areas along the Lake Erie shoreline, including the Fermi site (Figure 6). To prevent flooding of the developed areas, these areas were elevated during the construction of Fermi 2 using crushed limestone taken from the southwest portion of the Fermi site (Quarry Lakes). Site elevations range from the level of Lake Erie to approximately 25 feet above lake level on the western edge of the site (Reference 8). Topography on the Fermi site is relatively level in the undeveloped areas, with an elevation range of approximately 10 feet over the site according to U.S. Geological Service (USGS) topographic maps.

3.1.4 Soils

The overburden soils at the Fermi site consist of lacustrine deposits, glacial till, and rock fill (Figure 7). The rock fill is present only in the immediate area of the reactor; therefore, in the wetland areas, the overburden soils consist of lacustrine deposits and glacial till. The overburden is underlain by the Bass Islands Group dolomite bedrock. Groundwater is present in the overburden and the bedrock. The groundwater in the overburden is unconfined, while the Bass Islands Group aquifer is confined. The glacial till acts as an aquitard between the unconfined groundwater in the overburden and the confined groundwater in the Bass Islands Group aquifer.

The Monroe County Soil Survey (Reference 9) lists soil series Lenawee silty clay loam, ponded (Map Symbol 10) and Lenawee silty clay loam (21) as the primary mapped soil types on the Fermi site. Other soils found on the Fermi property include: urban land (63) on the eastern portion of the site where the

6

existing Fermi 1 and Fermi 2 buildings and infrastructure are located; urban land-Lenawee complex (57) on the southern edge of the Fermi site; Aquents complex (31) and Blount loam (13A) on the northwestern side of the site; Pits-Aquents complex (33) in the southeast portion of the site; water (W) primarily in the southeast and northeast portions of the site; and beaches (27) along the eastern edge of the Fermi property adjacent to Lake Erie. Figure 7 depicts the soil series identified.

3.1.5 Vegetative Communities

Vegetative communities and wetland habitats were evaluated during detailed terrestrial surveys conducted from 2008 through 2010. In 2008 and 2009, spring, summer and fall pedestrian surveys of flora and fauna were conducted in all habitat types including wetlands on the Fermi site (Reference 10). In 2010 individual wetlands were revisited to determine Michigan Natural Community classification and wetland condition and quality. Several upland and wetland vegetative communities have been distinguished at the Fermi site as listed in Section 3.1.2 - Land Use. An in-depth discussion of vegetative communities for wetland covertypes is provided in Section 3.1.8 - Wetlands.

Requests for data concerning known or potential occurrences of endangered, threatened, candidate, or special concern plant species on the Fermi site were submitted to the USFWS and the Michigan Natural Features Inventory. In addition, a list of threatened, endangered, or candidate species for Monroe County, Michigan was obtained online from the Michigan Natural Features Inventory. The American lotus (*Nelumbo lutea*) is a state threatened plant species. However, large local populations of American lotus are scattered in areas of southern Michigan, reaching an apparent peak in Monroe County (Reference 11). In the south lagoon, and to a lesser extent in the north lagoon, are large stands of American lotus. American lotus is also abundant in the South Canal (Figure 8).

3.1.6 Wildlife

As discussed in Section 3.1.5 and Section 3.1.8, the Fermi site includes several ecological communities, some of which are considered rare and imperiled. The Fermi site was extensively surveyed for wildlife in 1973 and 1974 (Reference 12) with updates to species occurrences in 2000 and 2002 as part of a wildlife habitat planning effort. The most recent terrestrial and aquatic wildlife surveys were conducted during 2008 and 2009 (References 13 and 14) to confirm data from earlier surveys and to further characterize the wildlife species using the Fermi property. Secondarily, the surveys aided in determining if important species use the site and to guide decisions concerning avoiding, minimizing or compensating for impacts to these species from the proposed expansion. As such, wildlife surveys focused on portions of the Fermi site where construction and operation of Fermi 3 could potentially impact wildlife, whether from habitat destruction, conversion to other habitat types or through general habitat degradation.

The USFWS was consulted concerning the occurrence or potential occurrence of species on or in the vicinity of the Fermi property that are protected under the Endangered Species Act. The USFWS stated that the project occurs within the potential range of some federally listed species, but that the USFWS

7

had no records of occurrence on the Fermi site or in the vicinity, nor was there any designated critical habitat in the area. The USFWS further stated that because of the types of habitat present at Fermi, no further action is required under Endangered Species Act. The USFWS did state that if more than 6 months pass before the project is initiated, then the USFWS should again be contacted to ensure there have been no regulatory changes. Detroit Edison will continue consultations with the USFWS per their recommendations.

The MDNR and the Michigan Natural Features Inventory (Reference 15) was consulted regarding the presence of known or potential occurrences of state-listed threatened or endangered species on the Fermi site. The only species in the USACE/MDEQ-regulated project areas is the Eastern fox snake (*Pantherophis gloydi*).

Based upon the review of the data collected in the terrestrial and aquatic surveys there were no occurrences of federally and/or state listed threatened or endangered species. Based on avian surveys conducted during 2006-2008, the bald eagle (*Haliaeetus leucocephalus*) is the only migratory species of note that has been observed on the Fermi site. None of the previously observed bald eagle nests were observed on the Fermi site as of January 2011. During 2008, while wetland surveys were being conducted, two fox snakes were observed on two separate occasions. In addition, fifteen separate sightings were made by Detroit Edison employees between 1990 and 2007 with 1-6 snakes identified on each occasion. In addition to minimizing wetland impacts, the fox snake's primary habitat, Detroit Edison has developed a mitigation plan which will be implemented to minimize the project's impact to the species.

3.1.7 Site Hydrology

Currently the hydrology of the area is influenced by the physical processes of Lake Erie. Lake Erie has a perfect seiche fetch. With a predominant southwest wind, specific locations on Lake Erie are susceptible to great fluctuations in water levels due to sustained winds pushing the lake water to the east, and then, as the winds subside, the water levelizes across the lake. This creates large waterless expanses followed quickly by water inundating creek and river mouths, resulting in a bathtub like "sloshing" effect. This creates unique opportunities for both plants and wildlife. Other local hydrological conditions are dictated by the Swan Creek.

Water is seasonally to permanently present throughout the majority of the Fermi site. Average annual precipitation is approximately 35 inches and generally well distributed throughout the year. The site receives direct, surface runoff from a 2,440 acre drainage basin with cropland, wetland and forest as the primary cover types. Surface water is received from Lake Erie during periods of high water and storm events.

The hydrology of the Fermi palustrine emergent (PEM) wetland areas is controlled almost entirely by the elevation of surface water in Swan Creek and Lake Erie. The surface water in Swan Creek and Lake Erie

8

is directly connected to the PEM areas on the Fermi site. Five sets of large-diameter culverts connect the majority of the inland PEM areas west of Doxy Road with the PEM areas that are directly connected with Swan Creek and Lake Erie. These culverts allow free flow of surface water throughout the interconnected PEM areas. Therefore, the surface water level in the majority of the PEM areas is directly controlled by the surface water elevation of Lake Erie and Swan Creek, rather than groundwater levels. Figure 9 shows the culvert locations and movement of surface water on the Fermi site.

Palustrine forested (PFO) and palustrine scrub-shrub (PSS) areas on the Fermi site are, for the most part, contiguous with the PEM areas. Therefore, these areas are hydraulically connected with the PEM wetlands, so the groundwater level in these areas is influenced by the surface water levels in Swan Creek and Lake Erie. With the exception of a few wetlands separated by berms or roads, the majority of wetland communities on the Fermi property are hydrologically connected and thus considered one wetland system.

3.1.8 Wetlands

Detroit Edison conducted assessments of wetland resources on 1,106 acres of undeveloped lands at the Proposed Development site between 2008 (Reference 16) and 2011. The purpose of these assessments is to identify and integrate natural resource considerations throughout the design and implementation phases of the Proposed Development and to guide mitigation measures including avoidance, minimization and the development of a high quality mitigation strategy to compensate for unavoidable impacts. The assessments are based on existing data and onsite data collection. Existing data include topographic maps, federal and state wetland maps, soil maps, aerial photos, land use data, and ecological survey data from previous studies. Onsite assessment data were collected in each year to delineate wetland boundaries, evaluate wetland functions and services, determine natural community types and assess wetland condition and quality. A jurisdictional determination was completed and minor edits to wetland boundaries were made in 2011 (Figure 10). Watershed assessments of the northern section of the Ottawa-Stony Creek watershed and the coastal zone of Western Lake Erie in Monroe County were completed to further inform development strategies and conservation priorities at the Proposed Development site. This section provides an overview of wetlands with potential impacts associated with the Proposed Development. Section 3.1.9 provides a summary of the watershed assessments.

A functional assessment based on the USACE New England Highway Method (Reference 17) was originally conducted during the 2008 field delineation (Reference 16). In 2010, field observations of wetlands with proposed impacts included a refined assessment of vegetation communities and other wetland characteristics to further describe the condition, functions and services of impact areas. Data collection and analysis methods were based on the Michigan Rapid Assessment Method for Wetlands (MiRAM, Reference 18) and the Delaware Rapid Assessment Procedure (Reference 19) and included metrics such as wetland size and connectivity, adjacent area use, hydrologic alterations and soil

9

disturbance, habitat structure, and presence of invasive species. The results of the 2008/2009 terrestrial surveys, 2010 field visits described above, and feedback from regulatory staff were used to further evaluate individual wetlands potentially impacted by the Proposed Development.

Over 500 acres of wetland were delineated at the Proposed Development site. The majority of wetlands at the Fermi site were ranked low to medium quality based on factors including hydrological disturbance, presence of invasive species, adjacent land use, fragmentation, human activity, deforestation, etc. There were several wetlands ranked high quality based on connectivity, presence of native, diverse vegetation communities, and wildlife habitat potential. Several other wetlands were given high ecological value based solely on their rare and imperiled status in Michigan even though condition ratings were low (MiRAM guidance, see below). Depending on condition, the principal functions and services provided by wetlands on the Fermi site include flood flow alteration, sediment/toxicant retention, nutrient removal, and fish and wildlife habitat.

Wetlands with proposed impacts and their associated covertypes are presented in Table 1. Mitigation is proposed for approximately 35.55 acres of potential impacts to regulated wetlands due to the Proposed Development. These potential impacts include approximately 10.90 acres of Great Lakes marsh, 3.15 acres of southern hardwood swamp, 3.91 acres of southern shrub carr, 0.80 acres of coastal emergent wetland, 10.53 acres of other emergent wetland, 4.89 acres of other forested wetland and 1.37 acres of other scrub shrub wetland.

3.1.9 Watershed Analysis

As part of the natural resource assessment effort, Detroit Edison conducted a watershed analysis to provide a broader geographic context to guide land use decisions at the Fermi site. The purpose of the watershed assessment is to provide an analysis of land use features of the inland and coastal watersheds that encompass the Fermi site and evaluate the connection between natural resources on the Fermi site and site-specific and watershed conservation priorities. The watershed assessment also provides a landscape level perspective useful in consideration of any land use changes, proposed impacts and proposed compensation strategies.

The Fermi site is located in the northern portion of the Ottawa-Stony watershed (OSW, Figure 11), USGS Cataloging Unit and Hydrologic Unit Code (HUC): 04100001 and the coastal zone of Western Lake Erie in Monroe County (CZM, Figure 12). The OSW drains areas to the north and west of Lake Erie and flows directly into the lake. The northern portion of the OSW has a drainage basin of approximately 182,733 acres and is dominated by agriculture (55%). Approximately 25% of the OSW land area is in natural cover and approximately 20% is developed (Figure 11). The CZM encompasses approximately 18,697 acres with an almost even interspersion of natural lands (38%), developed lands (38%) and agriculture (24%) (Figure 12). Protected lands for conservation and recreation make up approximately 4% of the OSW and 36% of the CZM.

10

Wetlands comprise approximately 6% of the OSW and 43% of the CZM. The OSW is dominated by vegetated wetlands. Forested wetlands comprise the majority of vegetated wetlands (60%) with the remainder being emergent (24%) and shrub/scrub (15%). The CZM has equal proportions of vegetated and non-vegetated (open water) wetlands. Emergent wetlands are the dominant type comprising 71% of the vegetated wetlands with the remaining wetlands being forested (17%) and scrub shrub (11%).

An approximation of historic wetlands for the OSW and the CZM was developed based on soils classified as >80% hydric (soils >80% of a soil map unit classified as hydric by the Natural Resources Conservation Service) and current mapped wetlands. Former wetlands were defined as areas that are mapped hydric soils (>80% of map unit) but not mapped as wetlands based on the latest wetland maps. The topography and landscape position of the OSW and CZM are ideal for the development of wetlands because the land is very flat and in close proximity to the coast of Lake Erie. Prior to European colonization, approximately 45% of the land area of the OSW was wetland (Figure 13). Based on the most recent wetland maps 6% of the OSW area is currently wetland which constitutes an 86% loss in the OSW. Historically, 77% of the land area of the CZM was wetland (Figure 14). Based on the most recent wetland maps, 43% of the CZM is wetland which constitutes a 44% loss in the CZM.

Watershed Conservation Priorities

Based on natural resource assessments conducted at the Fermi site and within the OSW and CZM, the following wetland-based conservation priorities were identified for this project:

1. Protect and restore existing high quality wetlands especially those that are directly connected to Lake Erie in the CZM and/or part of a larger wetland complex.

2. Improve a network of natural land use in the CZM and OSW by increasing the amount of large blocks (>50 acres) of natural lands and buffered streams to support ecosystem functions and services and establish corridors to connect large blocks.

3. Restore wetlands in the CZM to provide wildlife habitat and protect water quality in Lake Erie.

4. Restore wetlands and stream buffers in the OSW to re-establish large wetland complexes and riparian connections.

Because of the Fermi site's location in the lowest reaches of the OSW (in the CZM), any activity onsite will have the greatest local effects (either positive or negative) on coastal resources and Lake Erie itself. Based on the results of the watershed assessment, planned activities at Fermi have strategically avoided and minimized impacts to natural resources of high ecological value to the greatest extent possible. For unavoidable impacts, this mitigation strategy has been designed to address any loss of coastal habitat and the watershed conservation priorities listed above. Specifically, the proposed mitigation will restore approximately 130 acres of coastal wetland including Great Lakes marsh and southern hardwood swamp and reconnect this large block of natural land directly to Lake Erie via a restored and buffered stream channel. Approximately 21.4 acres of impacted wetlands will be restored post-construction on the Fermi

11

site. On- and offsite mitigation actions are in close proximity to existing conservation efforts to help establish connectivity and habitat corridors.

3.2 Mitigation Area

The following description of the mitigation area is based on field data and review of existing, available data including aerial photography, soil survey maps, USGS topographic maps, state and federal wetland mapping, Monroe County Drain Commissioner records, and as-built drawings for I-75. Field surveys were conducted for topography, soils, hydrology, and wetland communities between 2010 and 2012. Figure 15 provides a plan view of existing conditions including site boundary, surveyed topography, existing easements, and USACE Ordinary High Water Mark (OHWM). In Lake Erie, the OHWM extends approximately to the elevation contour of 573.4 feet referenced to the 1985 International Great Lakes Datum (IGLD 85).

3.2.1 Location and Ownership

The proposed offsite mitigation area, referred to as the Monroe site, is approximately 210 acres in size and 7.25 miles from the Fermi site on Detroit Edison's Monroe Plant, east of Interstate 75, north of La Plaisance Creek, immediately adjacent to Lake Erie (La Plaisance Bay), Town of Monroe, Monroe County, Michigan, in the Ottawa-Stony Watershed (HUC: 04100001, Figure 1). The mitigation site is owned and managed by Detroit Edison.

3.2.2 Land Use

The proposed mitigation targets a 173-acre agricultural field at the Monroe site (Figures 16 and 17). This portion of the site is currently farmed and includes small areas of remnant wetlands and dikes which separate the site from Lake Erie. Excess water is pumped from the fields to accommodate farming. Adjacent areas include a 36-acre conservation area with a wetland restored approximately 10 years ago and associated grassland buffer. Adjacent land uses also include active agriculture, early successional old field and shrub habitat, agricultural ditches, small forest patches, existing wetland habitat, industrial, residential and other developed areas, access roads, highways and Lake Erie. Historical maps and aerial photos indicate the land has been in agricultural use with no structures present.

3.2.3 Topography

The topography of the site is very flat with an average elevation of approximately 572 ft. Figure 15 provides surveyed elevations including OHWM as designated by USACE. The lowest elevations in existing ditches and swales are below 570 feet with the highest elevation located on the top of a small rise in the northwestern corner of the site at approximately 589 feet. The elevation of the dike separating the site from Lake Erie has an average elevation of approximately 578 feet. Average lake levels of Lake Erie are 571.5 feet with seasonal fluctuations and periodic seiches causing significantly higher and lower elevations.

12

3.2.4 Soils

The Monroe County Soil Survey soil mapping for the site shows the presence of two soil types within the site boundaries (Figure 18). These soil types include Warners silt loam and Lenawee silty clay loam. The Warners series consists of very deep, very poorly drained soils on nearly level floodplains and seepage areas of hillsides. The Lenawee series consists of very deep, very poorly drained soils in lacustrine deposits. These soils are on lake plains and in depressional areas on moraines, outwash plains, and glacial drainageways. Both mapped soils are hydric and suitable for wetland restoration/creation.

3.2.5 Vegetative/Wildlife Communities

Vegetative communities were observed at the mitigation site primarily during wetland delineation field visits. The dominant covertype is active agriculture (Figures 16 and 17). Other covertypes include a mix of wetlands such as emergent marsh, floodplain forest, southern shrub-carr and wet meadow, and uplands such as old field, successional shrub and forest. The MDNR and the Michigan Natural Features Inventory (Reference 15) was consulted regarding the presence of known or potential occurrences of state-listed threatened or endangered species on the mitigation site. Based on review of known or potential occurrences and observations during field data collection, there are no occurrences of federally and/or state listed threatened or endangered species at the site. The shallow waters of La Plaisance Bay, immediately adjacent to the site, support a population of American Lotus. Restoration of the site will likely provide additional habitat for this state-threatened species.

3.2.6 Site Hydrology

The mitigation site receives runoff from the 588-acre Davis Drain watershed. The Davis Drain, under the jurisdiction of the Monroe County Drain Commissioner, is located along the southwest corner of the site. The drain carries stormwater runoff from Interstate 75 and upstream property. Water is seasonally to permanently present in ditches, swales and small remnant wetlands on the project site. Average annual precipitation is 31.5 inches and generally well distributed throughout the year. The site receives direct runoff from a 250-acre drainage basin with cropland, wetland and forest as the primary covertypes. The hydrology of the site is influenced by extensive tile and ditching for the purpose of draining surface water to facilitate farming. Figure 19 illustrates the location of ditches, culverts, and direction of flow for surface water drainage. Excess water is pumped from the fields at the northeast corner of the site into the adjacent ash basin. There is currently no direct hydrological connection between the mitigation site and Lake Erie. Depth to groundwater has not been determined however soil borings up to 20 inches revealed a compact clay lens and no groundwater penetration; the mitigation site is primarily surface-water driven.

A hydrological study was conducted for the mitigation site and the drainage basin. A water budget was developed to support mitigation design. Two models were developed to estimate the average annual volume of water that could enter the mitigation site from the drainage basin and from the planned mitigation wetland itself. Models include estimates of peak flows and average rainfall volume of the Davis

13

Drain. Water budget calculations for the proposed wetland mitigation plan demonstrate the sustainability of the wetland design.

3.2.7 Existing Wetlands

The mitigation site is adjacent to and includes existing wetlands, some of which are mapped on USFWS National Wetland Inventory (NWI) maps as PFO, PSS and PEM wetland types (Figure 20). Wetland boundaries within the mitigation site were delineated in 2011 (Reference 38) and a jurisdictional determination was completed. A total of 13 wetlands areas (Figure 21) were identified on the site totaling 74.52 acres. These wetlands are distributed throughout the site with the greatest concentration adjacent to site drainage ditches and the near shore areas adjacent to the dike separating the site from Lake Erie. The majority of wetlands identified at the site are significantly impacted by ongoing agricultural activities including plowing and manipulation of site hydrology (draining). Low diversity and the presence of invasive species such as reed canary grass (*Phalaris arundinacea*) and common reed (*Phragmites australis*) are typical of many of these existing wetlands. A functional assessment and conditions assessment were conducted during wetland delineations using the same methods that were used at the impact site and described in Section 3.1.8. Eleven of the 13 wetlands (Wetlands 1-5, 7, 11-14, 16) were ranked low to medium quality based on factors including hydrological disturbance, presence of invasive species, adjacent land use, fragmentation, human activity (farming), deforestation and degree of departure from their original functions and values. Two wetlands (Wetlands 8 and 10) were assigned high ecological value based solely on their rare and imperiled status in Michigan even though condition ratings were low (MiRAM guidance). A description of individual wetlands is provided in Reference 38.

4.0 MITIGATION SITE SELECTION FACTORS

An extensive exploration of potential mitigation projects spanning several years both on- and offsite within the Ottawa-Stony Watershed and coastal zone of Western Lake Erie has been conducted. The offsite mitigation project proposed here was determined to be the best based on site selection factors including:

- location, size and attributes of existing habitat;
- quality of mitigation options and likelihood of success based on both ecological and economic factors;
- land ownership and availability;
- adjacent land use;
- value and proximity to existing conservation plans, projects and watershed priorities;
- connectivity of habitat types;
- possible benefits to threatened and endangered species; and
- stewardship capabilities.

The mitigation site is in the coastal zone of Lake Erie immediately adjacent to the lake. It is one of only a few existing restoration opportunities for rare and imperiled coastal wetlands along the western edge of

14

Lake Erie. This valuable restoration opportunity has the potential to provide habitat for threatened and endangered plant, fish and wildlife species that rely on this highly impacted habitat type. The mitigation site originally supported coastal wetland habitat. Agricultural activities resulted in ditching, draining and isolation from the lake by construction of a farm dike along the eastern edge of the property. In spite of drainage and ongoing agricultural activities at the site, the topography, soils and access to hydrology from both the lake and the upstream watershed remain typical of coastal wetland systems and supportive of restoration efforts. Once artificial drainage features are removed and the site is reconnected directly to Lake Erie, wetland functions will be restored with a high likelihood of success. The mitigation site is adjacent to an existing conservation area restored by Detroit Edison in partnership with USFWS.

Restoration of coastal wetlands is a priority conservation activity for natural resource agencies and organizations. The mitigation design integrates ecological attributes of coastal wetlands at the impact site and high quality wetlands managed by natural resource agencies along Western Lake Erie. These include direct connection to lake hydrology, establishment of microtopography, interspersion of wetland types, irregular shoreline, shallow slopes and habitat structures. The existing topography, soils and access to hydrology at the mitigation site support restoration of a diverse coastal wetland system that is ecologically responsive to Lake Erie water level fluctuations. Plantings will augment the existing natural wetland seed bank. These factors along with the resource capacity and commitment of Detroit Edison to protect and manage the wetland mitigation effort from design through long term management ensure a successful mitigation strategy.

5.0 MITIGATION WORK PLAN

Implementation of the mitigation plan will commence prior to or concurrent with wetland impacts at the Fermi site and once all necessary permits are in place. A plan set has been developed detailing the final design for the mitigation site including an overall site plan, grading plan and details, planting plan, and erosion and sediment control plan. Qualified contractors will be secured to construct mitigation elements and to provide professional oversight and management of project implementation. Measures as detailed in the invasive species management plan in Section 9.1 will be utilized to prevent the establishment of invasive species within the mitigation sites. All equipment brought to the site will be thoroughly cleaned of all soil before entry into any of the mitigation zones. All soil materials and amendments brought to the mitigation site from offsite locations will require pre-approval by the site inspector to ensure that these materials are not sources of potential invasive species contamination.

Mitigation design emphasizes heterogeneity in vegetation and hydrology to maximize ecological diversity and functional resilience of the wetland. Wetland restoration activities are designed to emphasize techniques that restore functions such as flood flow attenuation and storage, sediment/toxicant retention, nutrient removal, food chain support, breeding and migration habitat for migratory birds, breeding and over-wintering habitat for amphibians, increased nutrient cycling, increased connectivity of coastal habitat types, and water quality improvements for surface outflow. A natural buffer will be established or existing

15

buffers maintained to protect mitigation wetlands. This final mitigation design is based on a full site evaluation and has been developed in cooperation with existing conservation focus areas (e.g., Detroit River International Wildlife Refuge), watershed plans and priorities, and input from local, state and federal conservation agencies and organizations.

Wetland restoration efforts will replace and repair habitat modified by agricultural practices and hydrological disturbance within sensitive coastal areas. Mitigation actions will increase the abundance, integrity and quality of aquatic habitat types that are currently listed as rare and imperiled in the state of Michigan. The mitigation actions described below will restore wetlands in the 173-acre agricultural area as illustrated in Figure 3. The mitigation actions will include forested, scrub shrub, and emergent wetland (including open water and wet meadow wetland types) with direct hydrological connection to Lake Erie. A specific objective of the offsite mitigation area is to reestablish a direct connection between the current agricultural area and Lake Erie and to redirect runoff from Interstate 75 into the restored wetland. These actions will reconnect a relatively large coastal floodplain area and will allow water to be filtered before it reaches Lake Erie.

5.1 Construction and Planned Hydrology

Construction activities in the agricultural area will include clearing, excavating and grading the proposed mitigation area to target elevations conducive for development of Great Lakes marsh including open water and wet meadow zonation, southern hardwood swamp, and southern shrub-carr wetlands. The construction sequence is described in Section 5.3. The mitigation area will be restored to two separate but hydrologically connected wetland units. The eastern unit will be directly connected to Lake Erie via a 60-foot cut in the existing dike to an elevation of 569 feet. Water levels in the eastern unit will fluctuate with Lake Erie water levels. A meandering waterway with a bottom channel width of 60 feet and 10:1 side slopes will be excavated to the west of the lake connection to allow for a permanent open water marsh zone in the emergent marsh area, providing habitat for aquatic species. Several pools extending to an elevation of 567.5 feet connected by a narrow channel of similar elevation will be created within the meandering waterway in the eastern unit. Two of these pools nearest Lake Erie will be dug to approximately 563.5 feet to accommodate fish species overwinter and during times of extended low water. Grading of soils adjacent to this waterway including the development of a rolling, pit and mound topography, will provide for a variety of water levels and habitat types within the eastern unit.

The western unit will be connected to Lake Erie where the open water channel of the eastern unit meets the spillway and the water control structure controlling the western unit. The western unit is designed to have a more stable hydroperiod than the eastern unit. To achieve the desired wetland communities in the western unit, a low berm will be constructed between the eastern and western restoration units. This berm will be constructed to a top elevation of 575 feet with a 12-foot top width and 4:1 side slopes with armored sides to protect against erosion and muskrat activity. A spillway and water control structure will be set to a full service elevation of 574 feet. The water control structure will provide water level

16

management in increments of 6 inches from 574 feet to a complete drawdown. The berm, spillway and structure have been sized according to the drainage basin and hydrologic models to ensure adequate drainage capacity and successful restoration of proposed habitat types and acreages in the western unit. Additional hydrology will be introduced into the wetland by searching for and breaking drainage tile and plugging existing ditches. The western unit will be connected to the Davis Drain by allowing a small base flow to continue to Lake Erie and diverting a larger storm overflow to the wetland. This diversion will be accomplished by installing a small diameter culvert covered with soil in the Davis Drain. A cut in the Davis Drain bank upstream of this low flow culvert will be made to allow overflow to the wetland. This overflow will increase water flow into the wetland, slow floodwater, reduce sediment loading and filter toxicants from runoff water before it reaches Lake Erie.

Graded wetland basins (with the exception of open water channels) will integrate pit and mound topography and will be left rough to establish additional microtopography essential for creating niches for a variety of wetland plants. The edges of the excavated wetlands and transitions between wetland types will be irregular in shape with variable, shallow slopes.

5.2 Planned Vegetation and Habitat Features

5.2.1 Planned Vegetation

Recent surveys of the mitigation site have identified the presence of several invasive species, including common reed (*Phragmites australis*), reed canary grass (*Phalaris arundinacea*), flowering rush (*Butomus umbellatus*), and Canada thistle (*Cirsium arvense*). Purple loosestrife (*Lythrum salicaria*) has not been observed but is likely to occur in southeast Michigan in the habitat types present on the Monroe site. These species can be problematic if they are allowed to become established within mitigation areas. To ensure proper development of target vegetative communities, mechanical and chemical treatment of existing invasive species at the mitigation area will be conducted at least once before construction activities commence. Additional applications will be conducted if necessary. Response from native vegetation will be facilitated by removing dead, chemically treated vegetation through mechanical removal after each treatment. Section 9.1 below provides a detailed description of the Invasive Species Management Plan for the mitigation site pre- and post-construction.

Portions of the mitigation area that are currently farmed will be planted and seeded to establish native plant communities. Planting and seeding will also stabilize soil structure, provide biological diversity, restore ecosystem functionality, and protect against invasion by exotic and invasive herbaceous species. The constructed berm and all other upland construction areas will be seeded with a mix to prevent erosion, stabilize excavated areas and establish an herbaceous community typical of the region. Forested, shrub and emergent wetlands will be planted and seeded to closely resemble vegetation communities typical of southern hardwood swamps, southern shrub carr and Great Lakes marsh prior to invasion of common reed and other invasive and exotic species. These vegetation communities are described in Natural Communities of Michigan: Classification and Description (Reference 20).

17

A wetland seed bank is evident at the mitigation site and is expected to contribute to the development of target wetland communities. However, the primary method to establish target communities will be through direct seeding and planting. Seed and plant material will be from a recognized native seed and plant nursery and native to Michigan. A limited amount of hand collection of seed (up to 5% of seed requirement) may be conducted targeting key species from reference wetland locations or species that are not currently available from native nurseries. The genetic origin of all seed and plants will be from within 150 miles of the mitigation site to the maximum extent possible. A genetic origin within the eight-state Great Lakes region which includes Illinois, Indiana, Michigan, Ohio, Pennsylvania, Minnesota, New York and Wisconsin is also acceptable for species not commercially available with a genetic origin within a 150-mile radius. Wild-type nursery stock of an age and condition suitable for transplantation will be used. Seed will be applied in a manner and at a rate that will allow effective establishment of the wetland pool area and wetland margins. Seed distribution for adjacent wetland community types will be overlapped on slopes directly influenced by fluctuating lake levels to create a transitional zone that can respond to variable water regimes. These areas are typically dynamic in terms of plant and wildlife assemblages and exhibit high diversity. An overlapping seed distribution will support the development and responsiveness of these transition zones. Plant species are selected, and planting techniques will be applied, to emphasize both horizontal and vertical diversity of vegetation community structure. This aspect of the planting plan is supported by the grading plan that integrates microtopography including pits and mounds into all wetland community types.

Targeted species and associated details are provided by vegetation community type (Tables 3 through 7 and Figure 22). The Michigan Natural Features Inventory (Reference 20) for all target community types was used to create species lists. The Great Lakes marsh – emergent wetland was further refined to closely represent the common species found in this ecotype in Monroe County, MI (Reference 21). Plant species are chosen for their proven hardiness in the area, their ability to out-compete invasive plant species, wildlife value, availability, and their overall suitability to develop diverse, native communities. Individual plant species may be substituted with a native, ecologically similar species if the listed species are not available by the contracted seed/plant distributor at the time of implementation. Species in the planting plan tables are currently available from nurseries that are members of the Michigan Native Plant Producers Association (http://www.mnppa.org/members.html). Sources for plant materials include:

- The Native Plant Nursery LLC: http://www.nativeplant.com/
- Wildtype Plants- Mason, MI: http://www.wildtypeplants.com/
- Hidden Savanna Nursery : http://www.hiddensavanna.com
- Other MI native plant nurseries at: http://castle.eiu.edu/n_plants/michigan.htm

Seed will be purchased in quantities to support the overlapping seed distribution described above. Seed and plant quantities may be adjusted based on availability.

18

5.2.2 Habitat Structures

Habitat structures will be placed in all areas of the mitigation wetland with a grade of 570 feet or higher prior to seeding and planting. Habitat structures will be placed at a minimum of six per acre and include whole trees, logs, snags, tree stumps and sand mounds and are described in greater detail in Section 7, Item 2. Additional habitat structures in the form of snake and turtle hibernacula, basking and nesting structures may also be placed in appropriate locations on the mitigation site as directed by herpetological experts working with Detroit Edison on stewardship opportunities that will maximize the ecological value of the mitigation site beyond requirements for wetland compensation. These measures would augment the value of the proposed communities. They would not be in conflict with mitigation goals, objectives and performance standards.

5.3 Construction Sequence

The grading, planting, and introduction of hydrology at the offsite mitigation area will be constructed prior to or concurrent with initiating any Fermi 3 permitted activities. Construction is planned over a 4-year period to accommodate site preparation primarily in regards to eradicating existing invasive species and establishing planned hydrology. Invasive species control techniques will be applied in years 1 and 2 and each year thereafter, if necessary, as discussed in the Invasive Species Management Plan in Section 9.1. Farming is expected to continue until year 2 and assist in managing invasive plant species in the proposed mitigation area. The majority of the earthwork will be completed in year 2 along with seeding of all wetland community types and disturbed areas. Once seeded vegetation has been established in year 3, water levels on the west side of the wetland will be held to full service elevations and on the east side of the wetland the cut will be constructed to allow direct hydrological connection to Lake Erie. Water levels will be monitored throughout the rest of year 3 and into year 4. In year 4, plugs and container tree and shrub species will be installed. A summary of construction activities for each construction year and an approximate timeline is provided below.

- Year 1 - Initiate site preparation. Existing wetlands at the offsite mitigation area will be surveyed and treated with appropriate measures (manual removal and herbicide) to eradicate invasive plant species as described in the Invasive Species Management Plan in Section 9.1.
- Year 2 - Continue treatment of invasive plant species. Construction activities in the offsite mitigation area will include clearing, excavating and grading to elevations conducive for development of planned wetland communities. The berm separating the eastern and western units will be constructed and the water control structure and spillway will be installed along with the structure to allow flow from the Davis Drain onto the mitigation area. Habitat structures will be placed prior to seeding. Construction areas will be seeded with a mix to prevent erosion, stabilize excavated areas and establish an herbaceous community typical of the region.

19

Preconstruction meeting and site visit	June
Mobilization - install soil erosion control measures	June
Clearing and grubbing	June
Excavation and grading, construct berm, install water control structures	July - September
Install habitat structures	October
Final grading and seeding	October - November

- Year 3 – Manage western unit at full service water elevation. Excavate channel to connect the eastern unit of the mitigation site with Lake Erie.

Pre-Construction Meeting and Site Visit	June
Mobilization – install soil erosion control measures	June
Construct coffer dam	June
Excavate channel, install rip rap	July – August
Remove coffer dam	September
Remove spoils/Seed disturbed areas	October – November
Monitor water levels	November - May

- Year 4 – Complete final planting of plugs, tree/shrub potted materials after establishment of grade and hydrology. An assessment of water levels may require minor adjustments in grading to ensure proper hydroperiods are established for target wetland communities or minor adjustments in acreage goals for wetland community types.

Pre-construction meeting and site visit	June
Continue to monitor water levels	June - August
Adjust grade or hydrology, as required	August
Planting of potted nursery stock	October/May - June

20

6.0 PROTECTION

Ownership of on- and offsite mitigation areas will remain with Detroit Edison. The restored mitigation wetlands will be permanently protected as directed by regulatory requirements to preserve the wetland functions restored. Detroit Edison will execute a conservation easement over the mitigation area in a form identical to the conservation easement model on the MDEQ website at www.michigan.gov/deqwetlands. The original executed conservation easement and associated exhibits will be sent to the MDEQ for review and recording within 6 months of the Decision to Construct Fermi 3 and prior to commencing any permitted work within regulated areas. The boundary of the conservation easement is shown on Figure 23. The conservation easement boundary will be demarcated by the placement of signs along the perimeter. The signs will be placed at an adequate frequency, visibility, and height for viewing, made of a suitable material to withstand climatic conditions, and will be replaced as needed. The signs will include the following language:

WETLAND CONSERVATION EASEMENT

NO CONSTRUCTION OR PLACEMENT OF STRUCTURES ALLOWED.

NO MOWING, CUTTING, FILLING, DREDGING OR APPLICATION OF CHEMICALS ALLOWED.

MICHIGAN DEPARTMENT OF ENVIRONMENTAL QUALITY

7.0 PERFORMANCE STANDARDS

The following performance standards will be used to evaluate the mitigation wetland:

1. In the first monitoring year, a layer of high-quality topsoil, from the A horizon of an organic or loamy surface texture soil, is placed (or exists) over the entire wetland mitigation area at a minimum thickness of 6 inches.

2. In the first monitoring year, a minimum of six (6) habitat structures, consisting of at least three (3) types, have been placed per acre of mitigation wetland. At least 50 percent of each structure shall extend above the normal water level. This standard shall apply to all areas of the mitigation wetland with a grade of 570 feet or higher. The types of acceptable wildlife habitat structures are:

   a. Tree stumps laid horizontally within the wetland area. Acceptable stumps shall be a minimum of 6 feet long (log and root ball combined) and 12 inches in diameter.

   b. Logs laid horizontally within the wetland area. Acceptable logs shall be a minimum of 10 feet long and 6 inches in diameter.

   c. Whole trees laid horizontally within the wetland area. Acceptable whole trees shall have all of their fine structure left intact (i.e., not trimmed down to major branches for installation), be a minimum of 20 feet long (tree and root ball), and a minimum of 12 inches in diameter at breast height (DBH).

21

d. Snags which include whole trees left standing that are dead or dying, or live trees that will be flooded and die, or whole trees installed upright into the wetland. A variety of tree species should be used for the creation of snag habitat. Acceptable snags shall be a minimum of 20 feet tall (above the ground surface) and a minimum of 12 inches DBH. Snags should be grouped together to provide mutual functional support as nesting, feeding, and perching sites.

e. Sand mounds at least 18 inches in depth and placed so that they are surrounded by a minimum of 30 feet of water measuring at least 18 inches in depth. The sand mound shall have at least a 200 square foot area that is 18 inches above the projected high water level and oriented to receive maximum sunlight.

3. Planted woody species in the scrub-shrub and forested wetlands will achieve at least 70 percent survival one year after the site is planted. Survival is measured only during this establishment period. Any necessary replacement of dead woody plantings will ensure this performance measure is met.

4. Interim and final performance standards for the herbaceous layer mean percent cover of native hydrophytic species on the west and east sides of the constructed berm for each wetland type are as follows:

Year	Emergent	Wet Meadow	Shrub, Forested Wetlands
1	30	40	30
2	40	45	40
3	45	50	45
4	50	75	50
5	60 (Final)	80 (Final)	60
6 and 7			70
8 and 9			75
10			80 (Final)

The total percent cover of non-invasive, native, hydrophytic species in each plot shall be averaged for plots taken in the same wetland type to obtain a mean percent cover value for each wetland type. Plots within identified extensive open water and submergent areas, bare soil areas, and areas without a predominance of wetland vegetation shall not be included in this average. Hydrophytic species refers to species listed as facultative and wetter in the USACE 2012 National Wetland Plant List.

5. Interim and final performance standards for the minimum number of native hydrophytic plant species by wetland type are as follows:

22

Year	Emergent	Wet Meadow	Shrub, Forested Wetlands
1	7	7	7
2	8	10	8
3	10	12	10
4	12	15	12
5	15 (Final)	20 (Final)	15
6 and 7			15
8 and 9			15
10			15 (Final)

The total number of native hydrophytic plant species shall be determined by a sum of all species identified in sample plots of the same wetland type.

6. A Floristic Quality Assessment (Reference 23) will be conducted to evaluate plant community structure. The Floristic Quality Index including species richness and average conservatism of species will be calculated each monitoring year. The FQI of the mitigation site shall demonstrate a stable or increasing trend over the last two years of the monitoring period.

7. Interim and final performance standards for the number of individual surviving, established and free-to-grow trees per acre in the shrub and forested wetlands that are classified as native, hydrophytic wetland species and consisting of at least three different species are as follows.

At year 5 of the monitoring period, the mitigation wetland supports a minimum of:

    a. Two hundred (200) individual surviving, established, and free-to-grow trees per acre in the forested wetland that are classified as native wetland species and consisting of at least three different plant species.

    b. Two hundred (200) individual surviving, established, and free-to-grow shrubs per acre in the scrub-shrub wetland that are classified as native wetland species and consisting of at least four different plant species.

At the end of the monitoring period, the mitigation wetland supports a minimum of:

    c. Three hundred (300) individual surviving, established, and free-to-grow trees per acre in the forested wetland that are classified as native wetland species and consisting of at least three different plant species.

    d. Three hundred (300) individual surviving, established, and free-to-grow shrubs per acre in the scrub-shrub wetland that are classified as native wetland species and consisting of at least four different plant species.

8. Throughout the monitoring period the mean percent cover of invasive species including, but not limited to, *Phragmites australis* (Common Reed), *Lythrum salicaria* (Purple Loosestrife), and *Phalaris*

23

arundinacea (Reed Canary Grass) shall in combination be limited to no more than ten (10) percent within each wetland type. Invasive species shall not dominate the vegetation in any extensive area of the mitigation wetland.

If the mean percent cover of invasive species is more than ten (10) percent within any wetland type or if there are extensive areas of the mitigation wetland in which an invasive species is one of the dominant plant species, the permittee shall submit an evaluation of the problem to the USACE.

9. Extensive open water and submergent vegetation areas having no emergent and/or floating vegetation shall not exceed 20 percent of the mitigation wetland area west of the berm and 40 percent east of the berm.

10. By the end of the monitoring periods, extensive areas of bare soil shall not exceed five percent of the mitigation wetland area. For the purposes of these performance standards, extensive refers to areas greater than 0.01 acre (436 square feet) in size. The hydrologic variation experienced at this location will be considered when reviewing this standard.

11. At the end of the monitoring period, the mitigation wetland shall be free of oil, grease, debris, and all other contaminants.

12. At the end of the monitoring period the established wetlands will meet the federal wetland criteria outlined in the report entitled "Corps of Engineers Wetlands Delineation Manual" dated January 1987, as modified by all applicable supplements, associated lists, documents, etc. The site will be characterized by the presence of water at a frequency and duration sufficient to meet the hydrology criteria of the Corps of Engineers Wetlands Delineation Manual for at least three consecutive years and support a predominance of wetland vegetation and the wetland types specified. This will be documented in a final delineation report including a certified land survey of the wetland boundaries submitted to USACE prior to release of the mitigation.

If the mitigation wetland does not satisfactorily meet these final success criteria by the end of the monitoring period, or is not satisfactorily progressing according to interim success criteria during the monitoring period, the permittee will be required to evaluate and may be required to take corrective action.

This mitigation project was designed to replace functions and values of Great Lakes marsh by development of plant communities and zones as described in the Michigan Natural Features Inventory Natural Communities of Michigan: Classification and Description (Reference 20). This document recognizes that Great Lakes marshes are characterized by dynamic water level cycles that can dramatically alter vegetation zones and their placement on the landscape. Monitoring reports shall indicate if performance standards are not satisfactorily met due to these natural, dynamic hydrologic conditions with a description of corrective actions or an explanation if corrective actions are not merited.

24

8.0 MONITORING

Monitoring activities completed at the mitigation site will be conducted as described by MDEQ Technical Guidance for Wetland Mitigation represented below (Reference 1). This monitoring plan also satisfies USACE guidance contained in 33 CFR Part 332 – Compensatory Mitigation for Losses of Aquatic Resources. A monitoring plan is necessary to evaluate the mitigation wetland in regards to meeting the performance standards of the project. A biologist, experienced with wetland restoration and mitigation will coordinate and oversee monitoring activities. Detroit Edison will submit a surveyed drawing showing the as-built conditions of the mitigation area to MDEQ and USACE within 60 days following completion of construction. Monitoring visits will be performed annually beginning with the first growing season after construction is completed. Emergent wetlands will be monitored for a minimum of 5 years and shrub and forested wetlands will be monitored for a minimum of 10 years or until performance standards are met. Monitoring includes:

1. During construction provide one-time photographic documentation of high quality soil placement across the site.

2. Measure inundation and saturation at all staff gauges, monitoring wells, and other stationary points shown in the mitigation plan (Figure 24) monthly during the growing season. Hydrology data shall be measured and provided at sufficient sample points to accurately depict the water regime of each wetland type.

3. Sample vegetation in plots located along transects shown in the mitigation plan (Figure 24) once between July 15 and August 31 or other timeline required to adequately sample target vegetation communities. The final number of sample plots necessary within each wetland type shall be determined by use of a species-area curve. The minimum number of sample plots for each wetland type shall be no fewer than five (5). Sample plots shall be located on the sample transect at evenly spaced intervals. If additional or alternative sample transects are needed to sufficiently evaluate each wetland type, they must be approved in advance in writing by regulatory staff. The herbaceous layer (all non-woody plants and woody plants less than 3.2 feet in height) shall be sampled using a 3.28 foot by 3.28 foot (1 square meter) sample plot. The shrub and tree layer shall be sampled using a 30-foot radius sample plot. The data recorded for each herbaceous layer sample plot shall include a list of all living plant species, and an estimate of percent cover in 5 percent intervals for each species recorded, bare soil areas and open water relative to the total area of the plot. The number and species of surviving, established and free-to-grow trees and surviving, established, and free-to-grow shrubs shall be recorded for each 30-foot radius plot. Plot data and a list of all the plant species identified in the plots and otherwise observed during monitoring will be provided. Data for each plant species will include common name in English, scientific name, wetland indicator category from the USFWS's National List of Plant Species That Occur in Wetlands for Region 3 (Reference 22), whether the species is considered native according to the Michigan Floristic Quality Assessment

25

(Reference 23) and associated coefficient of conservatism value. Nomenclature shall follow Reference 24 through Reference 26. Data will be used to calculate diversity, species richness, mean coefficient of conservatism values and a Floristic Quality Index for the mitigation site. Water depth measurements will be taken at the center of each sampling plot. The location of sample transects and plots will be identified in the monitoring report on a plan view showing the location of wetland types. Sample transects shall be permanently staked at a frequency sufficient to relocate the transect in the field.

4. Delineate any extensive (greater than 0.01 acre in size) open water areas, bare soil areas, areas dominated by invasive species, and areas without a predominance of wetland vegetation, and provide their location on a plan view.

5. Document any sightings or evidence of wading birds, songbirds, waterfowl, amphibians, reptiles, and other animal use (lodges, nests, tracks, scat, etc.) noted within the wetland during monitoring. Note the number, type, date, and hour of the sightings and evidence.

6. Inspect the site during all monitoring visits and inspections for oil, grease, man-made debris, and all other contaminants and report findings. Rate (e.g., poor, fair, good, excellent) and describe the water clarity in the mitigation wetland and determine source(s) of turbidity.

7. Provide annual photographic documentation of mitigation wetland development during vegetation sampling from permanent photo stations located within the mitigation site. At a minimum, photo stations shall be located at both ends of each transect. Photos will be labeled with the location, date, and direction.

8. Provide the number, type and location of habitat structures placed and representative photographs of each structure type.

9. Conduct a wetland delineation to determine the area meeting all three wetland criteria (dominance by hydrophytic vegetation, wetland hydrology and hydric soils) at the completion of the monitoring period. Include the wetland delineation in the final monitoring report as a supplement and include the estimated wetland acreage in the report.

10. Provide a written summary of data from previous monitoring periods and a discussion of changes or trends based on all monitoring results.

11. Provide a written summary of all the problem areas that have been identified and potential corrective measures to address them.

Monitoring reports shall cover the period of January 1 through December 31 of each year following planting. Reports will be submitted to Detroit Edison before January 31 of the following year. Detroit Edison will forward the annual reports to the appropriate regulatory agencies. Additional monitoring

26

beyond the 5 or 10-year standard monitoring period may be required if all performance standards are not met to the satisfaction of MDEQ and USACE.

9.0 MAINTENANCE, ADAPTIVE MANAGEMENT AND INVASIVE SPECIES MANAGMENT

Necessary steps will be taken to ensure the proper establishment and maintenance of the mitigation wetland. The mitigation site will be visited one to two times each year by qualified contractors during the monitoring period to satisfy standard maintenance requirements and to identify any conditions that threaten the proper protection, function and development of the wetlands, streams and associated buffers. Any deficiencies in vegetative community development including plant survival will be noted and appropriate corrective measures will be implemented.

If monitoring indicates that a performance standard is not being met, that standard will be evaluated to determine if simply more time is needed or if a remedial action may be required. Remedial measures may include seeding or planting, non-native plant control, and erosion control measures. In less common circumstances contingency may be required regarding the wetland basin, removal or addition of dikes, spillways, or other water control structures, and access control. Should adaptive management be required, Detroit Edison will develop an adaptive management plan and implementation timetable and submit it to the MDEQ and USACE for review and approval. Upon approval, Detroit Edison will proceed with implementation of adaptive management activities.

9.1 Invasive Species Management Plan

Recent surveys of the mitigation site have identified the presence of several invasive species, including common reed (*Phragmites australis*), reed canary grass (*Phalaris arundinacea*), flowering rush (*Butomus umbellatus*), and Canada thistle (*Cirsium arvense*). Purple loosestrife (*Lythrum salicaria*) has not been observed but is likely to occur in southeast Michigan in the habitat types present on the Monroe site. These species can be problematic if they are allowed to become established within mitigation areas. Most of these species prefer wetland sites, but upland areas can be just as susceptible to colonization by some of these and other invasive species. These and most other invasive species produce many seeds, grow quickly, have few natural predators in the area, and can quickly produce monocultures within mitigation areas to the significant detriment of more desirable native species. The invasive species management program for the Monroe site includes measures to identify and address the presence of invasive species within the site boundary and adjacent areas owned by Detroit Edison.

Mechanical and chemical treatment of existing invasive species will be conducted at least once before construction activities commence. Additional applications will be conducted if necessary. One treatment should sufficiently control the existing invasive species to a point where they can effectively be monitored and treated during and after construction as necessary to minimize existing coverage of all onsite invasive species. Several existing wetlands and upland areas at the mitigation site will be treated with herbicide to kill invasive plant species including common reed, reed canary grass and Canada thistle prior to construction of the mitigation wetland. Response from native vegetation will be facilitated by removing

27

dead, chemically treated vegetation through burning or mowing after each treatment. Seeding and planting within the mitigation area will be conducted as soon as conditions allow following earthwork, limiting the potential for new infestations. After construction, the mitigation area will be monitored to allow for early detection of, and rapid response to, the future establishment of any invasive species.

9.1.1 Monitoring

Monitoring of the mitigation area has already begun with the preconstruction vegetation surveys and wetland delineation. Species present have been recorded and invasive species have been noted. Additional surveys will be conducted prior to construction activities to map the specific location of invasive species patches in preparation for control activities. Monitoring will be conducted using both visual ocular and transect surveys once after preconstruction treatment but before construction, monthly during construction, and semi-annually after construction activities have ceased, to identify any regrowth of original invasive patches as well as any colonization of new areas by invasive species. Post construction monitoring will continue annually through the life of the monitoring period. This monitoring will be conducted by Detroit Edison staff or a qualified contractor. Anyone involved with identification of invasive species will be given instruction in identification of all invasive species likely to occur in southeast Michigan in the habitat types present on the Monroe site. Emphasis will be given to those species present prior to construction. Estimates of the percent cover of invasive species will be based on qualitative ocular estimates and reported to MDEQ and USACE as part of the regularly scheduled monitoring reports. If invasive species are observed, they will be addressed in accordance with the following management procedures.

If the permittee determines that it is infeasible to reduce the cover of invasive species to meet the performance standard identified in Section 7, item 8, the permittee must submit an assessment of the problem, a control plan, and the projected percent cover that can be achieved for review by the USACE. Based on this information, the USACE may approve an alternative invasive species standard. Any alternative invasive species standard must be approved in writing by the USACE.

9.1.2 Invasive Plant Species Management

Invasive plant species most likely to be a problem in the restored wetland areas include common reed, purple loosestrife, reed canary grass and flowering rush. Additionally, upland areas within the site are likely to be degraded by the presence of Canada thistle. Each species is addressed below including a discussion of its ecology and control measures.

Common Reed (*Phragmites australis*)

Common reed is an aggressive grass with an extensive rhizome root system (http://plants.usda.gov/factsheet/pdf/fs_phau7.pdf). Once established, common reed can be extremely difficult to eliminate. While many control measures have been tried in the past, including mowing, flooding, burning, and covering with black plastic, the most effective control method has been herbicide

28

application. Glyphosate has been shown to be an effective control measure but may take two or three seasons of applications to eliminate dense stands. Other herbicides, such as Imazapyr, have recently shown promise in controlling common reed and may be an effective alternative to Glyphosate. MDEQ and Michigan Department of Natural Resources (MDNR), Ducks Unlimited, USFWS, and other participating land managers are currently experimenting with various techniques for controlling common reed in coastal wetlands along Lake Erie and Saginaw Bay. The techniques being tested include glyphosate, imazapyr, and a glyphosate/imazapyr mixture along with mechanical management actions. The treatment plan for existing and any future growth of common reed at the Monroe site is based on the MDEQ Guide to the Control and Management of Invasive Phragmites (Reference 27), any new, widely accepted, information resulting from Phragmites control studies, and on consultation with regulatory and conservation agency staff who have extensive knowledge of chemical control of invasive species in the coastal zone of Western Lake Erie.

Common reed is shade intolerant and once the planted shrub and forested species provide a canopy that shades the restoration areas, common reed should not be a concern. If common reed becomes established in the emergent marsh areas, it will remain indefinitely since no shading will be likely. Regardless of its location, common reed will be aggressively controlled on the entire mitigation site during the monitoring period. Hand pulling or digging may be effective on small or very young plants. This technique is very labor intensive particularly if the plant becomes well established. However, once a stand becomes established, the extensive root system will make hand pulling or digging very difficult and essentially ineffective. At this point the most effective means of control of common reed will be application of herbicides, usually glyphosate as discussed above.

Herbicide can be sprayed or applied by wick application. Glyphosate is a nonspecific herbicide and the foliage of any plant sprayed will be killed. Therefore, spraying will be conducted in a manner in which overspray of non-target species is minimized. Control of dense stands of common reed may require multiple applications over several years. Application of herbicide will be conducted using a concentration and during a time period that has been shown to be effective in southeastern Michigan (e.g., 6 pints/acre of Glyphosate sprayed in early September). Any herbicide application within the mitigation site will be conducted by a Michigan licensed herbicide applicator. Additionally, any herbicide sprayed within the wetland areas of the site will be approved for such applications.

Currently, several dense stands of common reed exist on the mitigation site. These stands total approximately 15 acres. These stands will be treated with ground application equipment at least once before construction activities commence. Additional applications will be conducted if necessary. One application should sufficiently control the existing common reed stands to a point where they can effectively be monitored and treated while construction activities are underway.

29

Purple Loosestrife (*Lythrum salicaria*)

Purple loosestrife is a wetland indicator species and often found in natural and man-made wetlands (http://plants.usda.gov/plantguide/pdf/pg_lysa2.pdf). This species can be effectively controlled by several methods. Typical control measures include hand pulling, herbicide treatment or biological control (*Galerucella* spp. beetles). Similar to common reed, purple loosestrife is shade intolerant and once the planted shrub and forested species provide a canopy that shades the restoration areas, purple loosestrife should not be a concern. If purple loosestrife becomes established in the emergent marsh areas, it will remain indefinitely without treatment since no shading will be likely.

Regardless of its location, purple loosestrife will be aggressively controlled on the entire mitigation site during the monitoring period. Young plants can be pulled up by hand or dug up if the plant is not too big and the infestation is not too widespread. This technique is very labor intensive particularly if the plant becomes well established. However, once a stand becomes established, the extensive root system will make hand pulling or digging very difficult and essentially ineffective. Once the plants get larger than 18 inches in height, or the density of plants is excessive, herbicide treatment with Glyphosate or another suitable herbicide, as described for common reed above, will be more effective to control purple loosestrife. Control of dense stands of purple loosestrife may require multiple applications over several years.

Biological control may provide the best opportunity for long term treatment of an extensive infestation of purple loosestrife. Control would be achieved by the release of two leaf-feeding species of *Galerucella* spp. beetles (*G. pusilla* and *G. calmariensis*). Adults and larvae of these species prefer purple loosestrife as a food source feeding on the leaves, significantly weakening the plants and can cause a reduction in purple loosestrife density of up to 90 percent. Biological control is not expected to completely eradicate purple loosestrife and utilizing this approach will require review of performance standards. Use of these beetles has been shown to be effective in controlling purple loosestrife in other locations in Michigan including the Fermi site. Michigan Sea Grant, a cooperative program of the University of Michigan and Michigan State University, and administered through the National Oceanic and Atmospheric Administration (NOAA), provides information on the efficacy and use of biological control for purple loosestrife in Michigan (http://www.miseagrant.umich.edu/ais/pp/index.html). Biological control will be applied as needed and coordinated with Michigan Sea Grant and appropriate regulatory staff.

To date, purple loosestrife has not been detected at the Monroe site.

Reed Canary Grass (*Phalaris arundinacea*)

Reed canary grass is an aggressive wetland species that forms dense monotypic stands to the exclusion of other wetland species (http://plants.usda.gov/factsheet/pdf/fs_phar3.pdf). It spreads by rhizomous growth and seeds. Once established it can be difficult to adequately control due to resprouting from the soil seed bank. Similar to the previously highlighted species reed canary grass is shade intolerant and

30

once the planted shrub and forested species provide a canopy that shades the restoration areas, reed canary grass should not be a concern. If reed canary grass becomes established in the emergent marsh areas, it will remain indefinitely without treatment since no shading will be likely. Some control may be realized by increasing water levels, but this could negatively affect desirable species as well. Regardless of its location, reed canary grass will be aggressively managed prior to construction and controlled on the entire mitigation site and adjacent areas owned by Detroit Edison where appropriate during the monitoring period.

Several methods of control are available each with moderate effectiveness. No one methodology will be fully effective if the reed canary grass is well established. Control methods include, herbicides, burning, mowing or mechanical removal. Use of Glyphosate has shown to have some success, being effective for up to two years. After two years, regrowth from the seed bank may reestablish the stand. Spraying large stands and or wicking small stands or individual plants will provide the best options. Repeated application will likely be needed. Burning and twice yearly mowing have also shown some success, but again resprouting from the seed bank will require management over multiple years. Removal using heavy construction equipment has not shown to be effective due to rapid regrowth from rhizomes and seeds left in the soil.

Currently, stands of reed canary grass are present in existing wetlands at the mitigation site.

Flowering Rush (*Butomus umbellatus*)

Flowering rush is a perennial aquatic herb that spreads via rhizomes (http://www.in.gov/dnr/files/FLOWERING_RUSH.pdf). It can grow as both an emergent along shorelines and as a submersed plant in rivers and lakes. Once established, it can form dense stands which crowd out native plants. It is difficult to identify, especially when not flowered, as it resembles many native emergent plants, including common bulrush.

Control methods include, cutting and hand digging of the plant. It is very difficult to eradicate with the use of herbicides, herbicides easily wash off the narrow leaves of the plant. Cutting the plant below the surface of the water is an effective method of control. Cutting will not kill the plant, however it will decrease the abundance. Several cuttings within the same growing season will be required. It is very important that all cuttings of the plant be removed, any cuttings left can re-sprout and cause further spread. Hand digging is also an option for isolated plants or small stands. Care must be taken to remove all root fragments. As with the cuttings, any disturbed root fragment left can re-sprout and lead to the spread of the plant. Raking and pulling of the plants are not recommended as methods for this reason. Once the plant is removed from the water it can still grow and spread, mainly through sending out new shoots from the root stalk. All plants and pieces removed should be thoroughly dried. Drying should not occur near a wetland or any body of water, large piles should be turned frequently to ensure adequate drying. Control methods will have to be continued as long as the plant is present on the site. There is a

31

small stand of flowering rush in a wetland adjacent to the mitigation site that will be treated prior to construction and monitored thereafter.

<u>Canada Thistle (*Cirsium arvense*)</u>

Canada thistle is an aggressive, creeping perennial weed that reproduces from vegetative buds in its root system and from seed (http://plants.usda.gov/java/profile?symbol=ciar4). Infestation generally occurs on disturbed soils. It is difficult to control due to its extensive root structure, which allows it to recover after control attempts.

The key to controlling Canada thistle is to stress the plant and force it to use stored root nutrients. It is able to recover from almost any control method due to these root nutrient stores. Successful control and eradication requires several years of action. There are several viable options for control, and the best management includes combining multiple methods. Grasses and alfalfa can effectively compete with Canada thistle. If desired, planting these species in areas with Canada thistle will aid in control. Herbicide control is also an effective method; however, it will need to occur for several years as described for common reed above. Mowing is another option for control, in conjunction with herbicide treatments. Mowing should occur on a monthly basis, over several growing seasons. This repeated mowing regime depletes nutrients stored in the roots of the plant. Control methods should continue as long the plant is a problem on the site.

Farmed wetlands and upland areas at the mitigation site are colonized by Canada thistle and will be treated before, during and after construction utilizing a combination of the methods described above.

<u>Control of Other Invasive Species</u>

It is possible that other invasive species, not discussed in this document may become established in the mitigation area. Monitoring activities will be conducted with identification of any new species infestations as a priority. If any new invasive species are observed during monitoring or other site activities, those species will be identified, the size of the infestation determined and the best control methods researched and implemented.

9.1.3 Summary of Invasive Species Control

This plan provides a number of potential management techniques for the most likely invasive species that will be encountered in this project. No single management technique may be adequate to address all invasive species problems. Monitoring will be conducted on the entire mitigation site, including all habitat types. Once established, invasive species can be very difficult to control and even harder to eliminate. Therefore, the most important component of this invasive species control program is early detection and rapid response to new invasive species infestations. If the presence of invasive species is noted, a response plan will quickly be prepared to address the problem and determine the most effective and efficient control program. Action will be taken as soon as conditions (e.g., weather, time of year, plant life stage, etc.) allow. If a new infestation moves beyond a few plants and into a large area of coverage, it is

32

likely that control will have to incorporate one or more techniques over multiple seasons. However, even under this circumstance, the most effective and efficient control techniques will be used in an effort to eliminate the problem as soon as possible. When determining the proper technique to use to control invasive species, many variables will be reviewed. Control techniques will be reviewed based on factors such as historical and recent research, range wide efficacy, local efficacy, ecological impact of the control technique, and onsite experience with the control technique.

Monitoring for invasive species will be conducted throughout the construction period as part of the regular construction environmental monitoring and will continue after completion of construction as part of the wetland mitigation monitoring. Results of invasive species monitoring and control measures will be reported in annual monitoring reports. The Long Term Management Plan will also incorporate periodic monitoring and management measures for invasive species as appropriate.

10.0 LONG TERM MANAGEMENT PLAN

As discussed in Section 6, ownership of the mitigation site will remain with Detroit Edison. The site will be permanently protected via a conservation easement. In addition, Detroit Edison will implement the following actions to ensure long term management for the mitigation site. The long term management actions will commence with the acceptance of the final mitigation monitoring report and regulatory approval that the mitigation site has met all necessary performance standards. Detroit Edison will commence long term management by developing all necessary stewardship agreements and endowments. Copies of agreements and documentation of endowment funds to support annual site visits and any necessary long term management actions will be provided to regulatory agencies for the permit file.

This long term management plan provides an overview of how the wetland mitigation site will be monitored and maintained after mitigation construction has been completed and final performance standards have been met. Detroit Edison will enter into a long term agreement with a suitable third party steward and establish an endowment to support third party review of site conditions and long term management activities. The responsibility of Detroit Edison and the third party steward is to implement the activities described here and to prescribe, execute and evaluate any necessary management actions.

The third party steward will be provided with a copy of the Final Aquatic Resource Mitigation Strategy and Final Design, which includes this long term management plan. Section 3.2 of the mitigation strategy provides detailed background on the mitigation site including location, site history, existing conditions and adjacent land use. Section 5 provides a detailed description of mitigation actions and community types targeted for development of the site. A copy of as-built conditions and detailed monitoring reports will also be provided to the third party steward to support and guide stewardship review and activities. Monitoring reports will include as-built conditions, a final wetland delineation identifying wetland community boundaries, documentation of any rare and imperiled vegetation communities and animal species, photo documentation, existing and potential threats and potential problem areas. The third party

33

steward will review all available information and conduct an initial site visit. Detroit Edison will establish permanent photo stations and water level monitoring stations designated for the long term management phase. Detroit Edison will conduct annual site visits to the mitigation site. During annual site visits qualified staff will:

- Traverse the perimeter of the mitigation site
- Traverse wetland areas including a representative sample of each wetland community type
- Take photos from permanent photos stations
- Collect water level data from permanent water level gauges
- Record anecdotal observation of plant and animal species
- Record observations of public use activities
- Record, photograph and map potential threats (invasive species, erosion, signs of incompatible public use, etc.)
- Record, photograph and map rare and imperiled communities/species
- Visit areas where threats were previously recorded and evaluate efficacy of previous management actions.
- Check perimeter signs demarcating the conservation easement boundary to ensure signs are in place and readable.

In addition to the items listed above, annual site visits will document adherence to the conservation easement ensuring there has been no alteration of topography, creation of unplanned paths, trails, or roads; placement of fill, dredging, or excavation; drainage of surface or groundwater; construction or placement of any structure; plowing, tilling, or cultivating the soils or vegetation; cutting, removal, or alteration of vegetation; including the planting of non-native plant species; construction of unauthorized utility or petroleum lines; storage or disposal of garbage, trash, debris, abandoned equipment; accumulation of machinery or other waste materials; use or storage of off-road vehicles; placement of billboards or signs; or the use of the wetland for the dumping of storm water.

An annual stewardship report will be submitted to the third party steward for review. This report wil include recommendations for any required management actions and a suggested implementation schedule and cost estimate. Management actions will be implemented at the appropriate time and for the appropriate duration. Management actions will be prescribed only in the case of a documented threat. Threats may include erosion, presence of invasive species, nuisance wildlife, changes to adjacent land use, incompatible use of wetland areas, missing or unreadable boundary signs. Recommended management actions may include:

- Water level manipulation
- Manual or chemical removal of undesirable plant species as described in the invasive species management plan in Section 9.1
- Control of nuisance wildlife

34

- Repairs to berm, spillway or water control structures as needed
- Water level management as needed to maintain healthy interspersion of water and emergent vegetation on the west side of the mitigation site.
- Monitoring and management of public use to ensure compatible activities.
- Water quality monitoring to protect from undesirable impacts from land use changes in adjacent areas.
- Clean up of trash and debris
- Repair and maintenance of conservation easement signs and designated public use trails and signage.

The annual stewardship report will also be used to inform and update the long term management plan to continue utilizing an adaptive management strategy for development and maintenance of the wetland communities at the mitigation site.

11.0 FINANCIAL ASSURANCES

Detroit Edison will provide financial assurances in the amount of $7,500,000 in the form of a letter of credit or bond to ensure that the replacement wetland is constructed, the conservation easement is recorded, monitoring is completed, and corrective actions are performed as required to comply with the mitigation requirements and conditions of MDEQ permit 10-58-0011-P. The financial assurance document shall be provided to and accepted by the MDEQ within 6 months after the Decision to Construct Fermi 3.

35

12.0 REFERENCES

1. Michigan Department of Environmental Quality Geological and Land Management Division. Technical Guidance for Wetland Mitigation. Available online at http://michigan.gov/documents/MDOT_Finalmitguidance_Wetland_(this_document_is_Part_of_A 7)_117907_7.pdf.

2. Bartoldus, C. C., E. W. Garbisch, and M. L. Kraus. 1994. Evaluation for Planned Wetlands (EPW). Environmental Concern Inc., St. Michaels, MD.

3. Not used.

4. Not used.

5. Not used.

6. Not used.

7. Michigan Department of Natural Resources, Michigan's Wildlife Action Plan, Southern Lower Peninsula. Available online at: http://www.michigan.gov/dnr/0,1607,7-153-10370_30909_31053-153463--,00.html, accessed October 4, 2007.

8. Detroit Edison, Enrico Fermi Atomic Power Plant Unit 2, Applicant's Environmental Report, Operating License Stage, Volume I, Supplement 4, February 1978.

9. U.S. Department of Agriculture, Natural Resources Conservation Service, Web Soil Survey, November 1981. Available online at http://websoilsurvey.nrcs.usda.gov/app/, accessed April 14, 2008.

10. Fermi 3 Terrestrial Vegetation Survey Final Report, Black & Veatch Corporation, November 2009.

11. Michigan State University Extension, Michigan Natural Features Inventory, Rare Species Explorer. Available online at: http://web4.msue.msu.edu/mnfi/explorer/index.cfm, accessed January 25, 2008.

12. 1973-74 Annual Report of the Terrestrial Ecological Studies at the Fermi Site, NUS Corporation, Ecological Sciences Department, Cyrus Wm. Rice Division, 1974.

13. Fermi 3 Extended Terrestrial Wildlife Survey Final Report, Black & Veatch Corporation, September 2009.

14. Aquatic Ecology Characterization Report, Detroit Edison Company Fermi 3 Project, Final Report, AECOM, November 2009.

15. Michigan State University Extension, Michigan Natural Features Inventory, Michigan's Special Animals, list effective April 9, 2009. Available online at: http://web4.msue.msu.edu/mnfi/data/specialanimals.cfm#grp, accessed January 5, 2011.

16. Ducks Unlimited, DTE Fermi II Site, Monroe County Wetland Investigation Report, April 2011.

36

17. U.S. Army Corps of Engineers 1995. The Highway Methodology Workbook Supplement. Wetland functions and value: A descriptive approach. U.S. Army Corps of Engineers New England Division. NEDEP-360-1-30a.

18. Michigan Department of Natural Resources and Environment, Michigan Rapid Assessment Method for Wetlands, MiRAM Version 2.1 User's Manual, July 23, 2010. Available online at: http://www.michigan.gov/deq/0,1607,7-135-3313_3687-240071--,00.html.

19. Jacobs, A.D. 2007. Delaware Rapid Assessment Procedure Version 5.1. Delaware Department of Natural Resources and Environmental Control, Dover, DE. 34 pp.

20. Kost, M.A., D.A. Albert, J.G. Cohen, B.S. Slaughter, R.K. Schillo, C.R. Weber, and K.A. Chapman. 2010. Natural Communities of Michigan: Classification and Description. Michigan Natural Features Inventory, Report No. 2007-21, Lansing, MI. 314 pp. Last updated 2010.

21. Albert, D.A, G. Reese, S.R. Crispin, M.R. Penskar, L.A. Wilsmann, and S.J. Ouwinga. 1988. A survey of Great Lakes marshes in the southern half of Michigan's lower peninsula. MNFI report for Land and Water Management Division of Michigan DNR, Coastal Zone Management Program (CZM Contract 10C-3). 116 pp.

22. Reed, Porter B. Jr. 1988. National List of Plant Species that Occur in Wetlands: North Central (Region 3). U.S. Fish and Wildlife Service, Washington D.C. Biol. Rept. 88(26.1). 112 pp.

23. Herman, K. D., L. A. Masters, M. R. Penskar, A. A. Reznicek, G. S. Wilhelm, W. W. Brodovich, and K. P. Gardiner. 2001. Floristic Quality Assessment with Wetland Categories and Examples of Computer Applications for the State of Michigan – Revised, 2nd Edition. Michigan Department of Natural Resources, Wildlife, Natural Heritage Program. Lansing, MI. 19 pp. + Appendices.

24. Voss, E.G. 1972. Michigan Flora. Part I, Gymnosperms and Monocots. Cranbrook Institute of Science, Ann Arbor, MI. 488 pp.

25. Voss, E.G. 1985. Michigan Flora. Part II, Dicots (Saururaceae—Cornaceae). Cranbrook Institute of Science, Ann Arbor, MI. 724 pp.

26. Voss, E.G. 1996. Michigan Flora. Part III, Dicots concluded (Pyrolaceae-Compositae). Cranbrook Institute of Science, Ann Arbor, MI. 622 pp.

27. Michigan Department of Environmental Quality - A Guide to the Management and Control of Invasive Phragmites, 2008. Available online at http://www.michigan.gov/documents/deq/deq-ogl-ais-guide-PhragBook-Email_212418_7.pdf.

28. ESRI, Aerial: World Imagery. Available online at: http://goto.arcgisonline.com/maps/World_Imagery, accessed October 2010.

29. Michigan Center for Geographic Information. U.S. Geological Survey Topographic map for Monroe County. Available online at http://www.mcgi.state.mi.us/mgdl/, accessed October 2010.

30. Michigan Center for Geographic Information. Natural Resources Conservation Service 2000 SSURGO Soil data: Soil Survey Geographic database for Monroe, Washtenaw, and Wayne

37

County, Michigan. Available online at: http://www.mcgi.state.mi.us/mgdl/, accessed November 2010.

31. U.S. Geological Survey, National Hydrography Dataset (NHD). Available online at http://nhd.usgs.gov, accessed December 2010.

32. Natural Resources Conservation Service, Land Use Land Cover – 2001, Data Available from U.S. Department of Agriculture GeoSpatial Data Gateway. Available online at http://datagateway.nrcs.usda.gov/GDGOrder.aspx?order=QuickState, accessed December 2010.

33. Michigan Department of Natural Resources and Environment Coastal Management Program. Coastal Zone: Michigan Department of Natural Resources and Environment email communications, September 29, 2010 and October 1, 2010.

34. Ducks Unlimited, GLARO GIS: Conservation and Recreation Lands (CARL), Existing Conservation Lands: Great_Lakes_CARL_20080228. Ducks Unlimited Great Lakes/Atlantic Regional Office (GLARO). Available online at: http://glaro.ducks.org/carl, accessed December 2010.

35. Ducks Unlimited, GLARO GIS: NWI Update Data, Draft Version for Washtenaw County: Washtenaw_MI_NWI_Current_Draft_01212008. Ducks Unlimited Great Lakes/Atlantic Regional Office (GLARO). Available online at: www.ducks.org/conservation/GLARO/3822/GISNWIData.html, accessed November 2010.

36. Michigan Center for Geographic Information. U.S. Fish and Wildlife Service 1979-1994. National Wetlands Inventory Data. Available online at: http://www.mcgi.state.mi.us/mgdl/, accessed December 2010.

37. Letter from Peter W. Smith (Detroit Edison) to Colette Luff (USACE), "Fermi 3 Offsite Mitigation Area Functional Evaluation, Supplementing Detroit Edison's Joint Permit Application (File Number 10-58-011-P)," 2011-MEP-F3COLA-0092, December 9, 2011.

38. Letter from Randall Westmoreland (Detroit Edison) to Colette Luff (USACE), "Fermi 3 Offsite Mitigation Area Wetland Delineation Report," 2011-MEPF3COLA-0054, July 28, 2011.

38

TABLES AND FIGURES

Table 1. Wetland Impacts and Attributes Summary Table (Sheet 1 of 2)

ID	Type/General Description	Total Size (acres)	Impact (acres)	Jurisdiction	Condition/ Primary Function
B	Linear PFO	0.76	0.76	MDEQ/USACE	Low/ Floodflow alteration, sediment, toxicant retention, nutrient removal and wildlife habitat
C	Great Lakes marsh, fragmented from Lake Erie by access roads, but connected hydrologically through culverts	48.18	9.73[a]	MDEQ/USACE	Medium (high ecological value)/ Floodflow alteration, sediment, toxicant retention, nutrient removal and wildlife habitat
D	Palustrine forested wetland with partially open canopy	1.37	1.37	MDEQ/USACE	Medium/ Floodflow alteration, sediment, toxicant retention, nutrient removal and wildlife habitat
E-North	North: Palustrine mix of scrub-shrub, emergent marsh/wet meadow, in two sections split by Wetland D,	2.67	1.87	MDEQ/USACE	Medium/Floodflow alteration, sediment, toxicant retention, nutrient removal and wildlife habitat for both portions of E
E-South	South: Southern shrub carr or other coastal wetland type	2.04	2.04	MDEQ/USACE	
F	PFO southern hardwood swamp, relatively intact,	31.07	2.71	MDEQ/USACE	Medium (high ecological value)/Floodflow alteration, sediment, toxicant retention, nutrient removal and wildlife habitat
H	PEM edge around a created open water pit	1.96	1.96	MDEQ	Low/Minimal floodflow alteration, sediment/toxicant retention and nutrient removal
I	PFO southern hardwood swamp, relatively intact, indirectly connected to Lake Erie, provides a buffer for the interior and less disturbed wetland	39.74	0.44	MDEQ/USACE	Medium (high ecological value)/Floodflow alteration, sediment, toxicant retention, nutrient removal and wildlife habitat
U	PEM edge around a created open water canal	3.46	3.46	MDEQ/USACE	Low/Minimal floodflow alteration, sediment/toxicant retention and nutrient removal.
W	PEM wet meadow wetland	4.59	4.59	MDEQ	Low/ Floodflow alteration, sediment, toxicant retention, nutrient removal and marginal wildlife habitat

39

Table 1. Wetland Impacts and Attributes Summary Table (Sheet 2 of 2)

ID	Type/General Description	Total Size (acres)	Impact (acres)	Jurisdiction	Condition/ Primary Function
Y	PFO fragmented early successional with mixed vegetation and a partially open canopy	1.14	1.14	MDEQ	Low/Marginal wildlife habitat for edge species and limited water storage.
AA	PEM established spoil area	0.80	0.80	MDEQ/USACE	Low/Minimal floodflow alteration, sediment/toxicant retention and nutrient removal
II	PEM ditch, contains vegetation communities with high structural diversity and low species diversity with well-established invasive species populations	0.52	0.52	MDEQ	Low minimal floodflow alteration, sediment/toxicant retention and nutrient removal
JJ	PSS established spoil area	1.37	1.37	MDEQ	Low/ minimal floodflow alteration, sediment/toxicant retention and nutrient removal
KK	PFO linear wetland, connected to the South Canal	1.62	1.62	MDEQ/USACE	Low/ floodflow alteration, sediment/toxicant retention, nutrient removal, marginal wildlife habitat for edge species
South Canal	PEM Great Lakes marsh hydrologically connected to Lake Erie	1.97	1.17	MDEQ/USACE	Medium/ fish and wildlife habitat, floodflow alteration, sediment, toxicant retention and nutrient removal

a. 2.29 acres of temporary impact associated with transmission line construction are excluded. The area will be restored immediately after construction and does not require additional mitigation as per regulatory guidance.

40

Table 2. Wetland Impacts and Proposed Mitigation

Wetland Type	Fermi 3 Impacted Areas (Acres)[a]	USACE Jurisdictional Impacted Areas (Acres)[a]	Proposed Mitigation (Acres)
Emergent Marsh			
Great Lakes marsh (rare/imperiled)	9.73	9.73	
Palustrine emergent (coastal)	0.80	0.80	
Palustrine emergent (other)	5.11	0	
Emergent Marsh Totals	15.64	10.53	69.99
Open water - Great Lakes marsh (rare/imperiled)	1.17	1.17	
Open water - emergent (other)	5.42	3.46	
Open Water Totals	6.59	4.63	27.25
Forested Wetland			
Southern hardwood swamp (rare/imperiled)	3.15	3.15	
Palustrine forested (coastal and other)	4.89	3.75	
Forested Wetland Totals	8.04	6.90	22.30
Scrub Shrub Wetland			
Southern shrub carr (coastal)	3.91	3.91	
Palustrine scrub shrub (other)	1.37	0	
Shrub/Scrub Wetland Totals	5.28	3.91	10.61
Wetland Totals	35.55	25.97	130.15

a. 2.29 acres of temporary impact associated with transmission line construction will be restored immediately after construction and does not require additional mitigation as per regulatory guidance.

41

Table 3. Great Lakes Marsh – Emergent Planting Plan

Great Lakes Marsh	67.69 acres		
Seed Mix Species List	Seeding Rate: 6 lbs/acre		
Common Name	Scientific Name	Form[a]	% by Seeds
Sweet flag	*Acorus calamus*	Seed/Plug	0.31
Common water plantain	*Alisma subcordatum*	Seed/Plug	2.81
Swamp milkweed	*Asclepias incarnata*	Seed/Plug	0.23
Swamp aster	*Aster puniceus*	Seed/Plug	0.38
American slough grass	*Beckmannia syzigache*	Seed	3.28
Nodding bur marigold	*Bidens cernua*	Seed	2.95
Bristly sedge	*Carex comosa*	Seed/Plug	1.41
Bottlebrush sedge	*Carex hystericina*	Seed/Plug	1.13
Awlfruit sedge	*Carex stipata*	Seed/Plug	1.59
Fox sedge	*Carex vulpinoidea*	Seed/Plug	1.88
Joe pye weed	*Eupatorium maculatum*	Seed/Plug	0.45
Common boneset	*Eupatorium perfoliatum*	Seed/Plug	0.75
Canada manna grass	*Glyceria canadensis*	Seed	3.47
Reed manna grass	*Glyceria grandis*	Seed	3.75
Southern blue flag	*Iris virginica*	Seed/Plug	0.09
Soft rush	*Juncus effusus*	Seed/Plug	4.69
Cardinal flower	*Lobelia cardinalis*	Seed/Plug	1.88
Great blue lobelia	*Lobelia siphilitica*	Seed/Plug	2.34
Monkey flower	*Mimulus ringens*	Seed/Plug	21.57
Pennsylvania smartweed	*Polygonum pennsylvanicum*	Seed	1.22
Pickerel weed	*Pontederia cordata*	Seed/Plug	0.03
Common arrowhead	*Sagittaria latifolia*	Seed/Plug	0.29
Dark green bulrush	*Scirpus atrovirens*	Seed	21.57
Soft-stem bulrush	*Scirpus validus*	Seed	4.36
Common bur reed	*Sparganium eurycarpum*	Seed/Plug	0.14
Blue vervain	*Verbena hastata*	Seed/Plug	17.44

a. Plugs will be planted at a density of 500 plugs/acre along open water emergent marsh transition zones comprised of a mix of the listed species where Seed/Plug is indicated in the Form column.

42

Table 4. Southern Wet Meadow – Emergent Planting Plan (Sheet 1 of 2)

Southern Wet Meadow	15.87 acres		
Seed Mix Species List	Seeding Rate: 6 lbs/acre		
Common Name	Scientific Name	Form	% by Seeds
Swamp milkweed	*Asclepias incarnata*	Seed	0.12
Eastern lined aster	*Aster lanceolatus*	Seed	7.58
Side flowering aster	*Aster lateriflorus*	Seed	0.6
Swamp aster	*Aster puniceus*	Seed	7.73
Blue joint grass	*Calamagrostis canadensis*	Seed	13.53
Marsh bellflower	*Campanula americana*	Seed	0.82
Fringed sedge	*Carex crinita*	Seed	0.56
Bottlebrush sedge	*Carex hystericina*	Seed	1.09
Hairy sedge	*Carex lacustris*	Seed	0.06
Wollyfruit sedge	*Carex lasiocarpa*	Seed	0.03
Shallow sedge	*Carex lurida*	Seed	0.29
Fen panicled sedge	*Carex prairea*	Seed	2.03
Sartwell's sedge	*Carex sartwellii*	Seed	0.16
Awlfruit sedge	*Carex stipata*	Seed	0.82
Upright sedge	*Carex stricta*	Seed	0.13
Water hemlock	*Cicuta maculata*	Seed	0.29
Swamp thistle	*Cirsium muticum*	Seed	0.02
Spike rush	*Eleocharis calva*	Seed	8.7
Joe pye weed	*Eupatorium maculatum*	Seed	2.3
Common boneset	*Eupatorium perfoliatum*	Seed	15.46
Northern bedstraw	*Galium boreale*	Seed	0.17
Fowl manna grass	*Glyceria striata*	Seed	15.46
Marsh St.John's wort	*Hypericum virginicum*	Seed	0.56
Jewelweed	*Impatiens capensis*	Seed	0.01
Southern blue flag	*Iris virginica*	Seed	0.02
Marsh pea	*Lathyrus venosus*	Seed	0.01
Water horehound	*Lycopus americanus*	Seed	12.56
Prairie loosestrife	*Lysimachia quadriflora*	Seed	0.22
Wild mint	*Mentha arvensis*	Seed	1.45
Marsh wild timothy	*Muhlenbergia glomerata*	Seed	0.54
Water smartweed	*Polygonum amphibium*	Seed	0.01

43

Table 4. Southern Wet Meadow – Emergent Planting Plan (Sheet 2 of 2)

Southern Wet Meadow	15.87 acres		
Seed Mix Species List	**Seeding Rate: 6 lbs/acre**		
Common Name	Scientific Name	Form	% by Seeds
Mountain mint	*Pycnanthemum virginianum*	Seed	1.06
Great water dock	*Rumex orbiculatus*	Seed	0.02
Common arrowhead	*Sagittaria latifolia*	Seed	1.47
Mad dog skullcap	*Scutellaria lateriflora*	Seed	0.16
Late goldenrod	*Solidago gigantea*	Seed	0.6
Swamp goldenrod	*Solidago patula*	Seed	0.87
Rough goldenrod	*Solidago rugosa*	Seed	2.23
Purple meadow rue	*Thalictrum dasycarpum*	Seed	0.27

44

Appendix K

Table 5. Southern Shrub-Carr – Shrub Wetland Planting Plan (Sheet 1 of 2)

Southern Shrub-Carr	10.84 acres				
Container Species					
Common Name	Scientific Name	Form	Size	Spacing	%
Black chokeberry	*Aronia prunifolia*	Flat/Cont	1 gal	10'x10'	5
Bog birch	*Betula pumila*	Flat/Cont	1 gal	10'x10'	15
Silky dogwood	*Cornus amomum*	Flat/Cont	1 gal	10'x10'	15
Red osier dogwood	*Cornus sericea*	Flat/Cont	1 gal	10'x10'	10
American hazelnut	*Corylus americana*	Cont	1 gal	10'x10'	5
Winterberry	*Ilex verticillata*	Cont	1 gal	10'x10'	10
Swamp rose	*Rosa palustris*	Flat/Cont	1 gal	10'x10'	5
Pussy willow	*Salix discolor*	Flat/Cont	1 gal	10'x10'	10
Elderberry	*Sambuscus canadensis*	Flat/Cont	1 gal	10'x10'	10
Meadowsweet	*Spiraea alba*	Flat/Cont	1 gal	10'x10'	5
Nannyberry	*Viburnum lentago*	Cont	1 gal	10'x10'	5
Shrubby cinquefoil	*Potentilla fruticosa*	Flat	1 gal	10'x10'	5
		TOTAL PLANTS		4,336	100

45

Table 5. Southern Shrub-Carr – Shrub Wetland Planting Plan (Sheet 2 of 2)

Southern Shrub-Carr	10.84 acres		
Seed Mix Species List	Seeding Rate: 6 lbs/acre		
Common Name	Scientific Name	Form	% by Seeds
Water plantain	Alisma subcordatum	Seed	4.17
Swamp milkweed	Asclepias incarnata	Seed	0.67
Blue joint grass	Calamagrostis canadensis	Seed	19.46
Tall bellflower	Campanula americana	Seed	2.95
Longhair sedge	Carex comosa	Seed	2.09
Bottlebrush sedge	Carex hystericina	Seed	2.09
Hairy sedge	Carex lacustris	Seed	0.09
Upright sedge	Carex stricta	Seed	0.18
Fox sedge	Carex vulpinoidea	Seed	8.69
Water hemlock	Cicuta maculata	Seed	0.42
Common boneset	Eupatorium perfoliatum	Seed	11.12
Northern bedstraw	Gallium boreale	Seed	0.24
Rattlesnake grass	Glyceria canadensis	Seed	10.29
Soft rush	Juncus effusus	Seed	6.95
Water horehound	Lycopus americanus	Seed	6.78
Dark green bulrush	Scirpus atrovirens	Seed	6.39
Wool grass	Scirpus cyperinus	Seed	11.82
Rufous bulrush	Scirpus pendulus	Seed	1.31
Softstem bulrush	Scirpus validus	Seed	1.08
Rough goldenrod	Solidago rugosa	Seed	3.21

46

Table 6. Southern Hardwood Swamp – Forested Wetland Planting Plan (Sheet 1 of 2)

Southern Hardwood Swamp	25.69 acres				
Container Species					
Common Name	**Scientific Name**	**Form**	**Size**	**Spacing**	**%**
Red maple	*Acer rubrum*	Cont	1 gal	10'x10'	5
Silver maple	*Acer saccharinum*	Flat/Cont	1 gal	10'x10'	15
Yellow birch	*Betula alleghaniensis*	Flat/Cont	1 gal	10'x10'	10
Tamarack	*Larix laricina*	Cont	1 gal	10'x10'	5
Eastern cottonwood	*Populus deltoides*	Cont	1 gal	10'x10'	5
Swamp white oak	*Quercus bicolor*	Cont	1 gal	10'x10'	10
Pin Oak	*Quercus palustris*	Cont	1 gal	10'x10'	5
Musclewood	*Carpinus caroliniana*	Cont	1 gal	10'x10'	5
Shagbark hickory	*Carya ovata*	Cont	1 gal	10'x10'	10
Hackberry	*Celtis occidentalis*	Cont	1 gal	10'x10'	2
Buttonbush	*Cephalanthus occidentalis*	Flat/Cont	1 gal	10'x10'	2
Gray dogwood	*Cornus racemosa*	Cont	1 gal	10'x10'	5
Running strawberry bush	*Euonymus obovatus*	Cont	1 gal	10'x10'	2
Michigan holly	*Ilex verticillata*	Cont	1 gal	10'x10'	5
Spicebush	*Lindera benzoin*	Cont	1 gal	10'x10'	5
Chokecherry	*Prunus virginiana*	Cont	1 gal	10'x10'	2
Wild black currant	*Ribes americanum*	Cont	1 gal	10'x10'	1
Swamp rose	*Rosa palustris*	Flat/Cont	1 gal	10'x10'	2
Elderberry	*Sambuscus canadensis*	Flat/Cont	1 gal	10'x10'	2
Nannyberry	*Viburnum lentago*	Cont	1 gal	10'x10'	2
		TOTAL PLANTS		10,276	100

47

Table 6. Southern Hardwood Swamp – Forested Wetland Planting Plan (Sheet 2 of 2)

Southern Hardwood Swamp	25.69 acres		
Seed Mix Species List	Seeding Rate: 6 lbs/acre		
Common Name	Scientific Name	Form	% by Seeds
Water plantain	Alisma subcordatum	Seed	4.17
Swamp milkweed	Asclepias incarnata	Seed	0.67
Blue joint grass	Calamagrostis canadensis	Seed	19.46
Tall bellflower	Campanula americana	Seed	2.95
Longhair sedge	Carex comosa	Seed	2.09
Bottlebrush sedge	Carex hystericina	Seed	2.09
Hairy sedge	Carex lacustris	Seed	0.09
Upright sedge	Carex stricta	Seed	0.18
Fox sedge	Carex vulpinoidea	Seed	8.69
Water hemlock	Cicuta maculata	Seed	0.42
Common boneset	Eupatorium perfoliatum	Seed	11.12
Northern bedstraw	Gallium boreale	Seed	0.24
Rattlesnake grass	Glyceria canadensis	Seed	10.29
Soft rush	Juncus effusus	Seed	6.95
Water horehound	Lycopus americanus	Seed	6.78
Dark green bulrush	Scirpus atrovirens	Seed	6.39
Wool grass	Scirpus cyperinus	Seed	11.82
Rufous bulrush	Scirpus pendulus	Seed	1.31
Softstem bulrush	Scirpus validus	Seed	1.08
Rough goldenrod	Solidago rugosa	Seed	3.21

48

Table 7. Mesic Southern Forest – Upland Planting Plan (Sheet 1 of 2)

Mesic Southern Forest	13.31 acres				
Container Species					
Common Name	Scientific Name	Form	Size	Spacing	%
Red maple	Acer rubrum	Cont	1 gal	30'x30'	10.0
Sugar maple	Acer saccharum	Flat/Cont	1 gal	30'x30'	20.0
Bitternut hickory	Carya cordiformis	Flat/Cont	1 gal	30'x30'	12.5
American beech	Fagus grandifolia	Cont	1 gal	30'x30'	12.5
Tulip tree	Liriodendron tulipifera	Cont	1 gal	30'x30'	7.5
Black cherry	Prunus serotina	Cont	1 gal	30'x30'	7.5
White oak	Quercus alba	Cont	1 gal	30'x30'	5.0
Northern red oak	Quercus rubra	Cont	1 gal	30'x30'	5.0
American basswood	Tilia americana	Cont	1 gal	30'x30'	5.0
Pawpaw	Asimina triloba	Cont	1 gal	30'x30'	2.0
Musclewood	Carpinus caroliniana	Flat/Cont	1 gal	30'x30'	2.0
Alternate-leaved dogwood	Cornus alternifolia	Cont	1 gal	30'x30'	2.0
Witch hazel	Hamamelis virginiana	Cont	1 gal	30'x30'	2.0
Spicebush	Lindera benzoin	Cont	1 gal	30'x30'	3.0
Virginia creeper	Parthenocissus quinquefolia	Cont	1 gal	30'x30'	2.0
Maple-leaf viburnum	Viburnum acerifolium	Cont	1 gal	30'x30'	2.0
		TOTAL PLANTS		644	100.0

49

Table 7. Mesic Southern Forest – Upland Planting Plan (Sheet 2 of 2)

Mesic Southern Forest	13.31 acres		
Seed Mix Species List	Seeding Rate: 7 lbs/acre		
Common Name	Scientific Name	Form	% by Weight
Big bluestem	Andropogon gerardii	Seed	8.93
Common milkweed	Asclepias syriaca	Seed	0.9
Butterfly milkweed	Asclepias tuberosa	Seed	0.45
Arrow-leaved aster	Aster sagittifolius	Seed	1.34
White wild indigo	Baptisia lactea	Seed	0.9
Partridge pea	Cassia fasciculata	Seed	3.93
Lance-leaf coreopsis	Coreopsis lanceolata	Seed	1.8
Purple coneflower	Echinacea purpurea	Seed	3.57
Canada wild rye	Elymus canadensis	See	28.57
Rattlesnake master	Eryngium yuccifolium	Seed	0.9
False sunflower	Heliopsis helianthoides	Seed	3.57
Wild bergamot	Monarda fistulosa	Seed	0.27
Switchgrass	Panicum virgatum	Seed	7.14
Foxglove beardtongue	Penstemon digitalis	Seed	1.8
Yellow coneflower	Ratibida pinnata	Seed	2.68
Black-eyed susan	Rudbeckia hirta	Seed	4.46
Brown-eyed susan	Rudbeckia triloba	Seed	0.27
Little bluestem	Schizachyrium scoparium	Seed	8.93
Indian grass	Sorghastrum nutans	Seed	17.86
Hoary vervain	Verbena stricta	Seed	1.8

50

Figure 1. Site Location Map

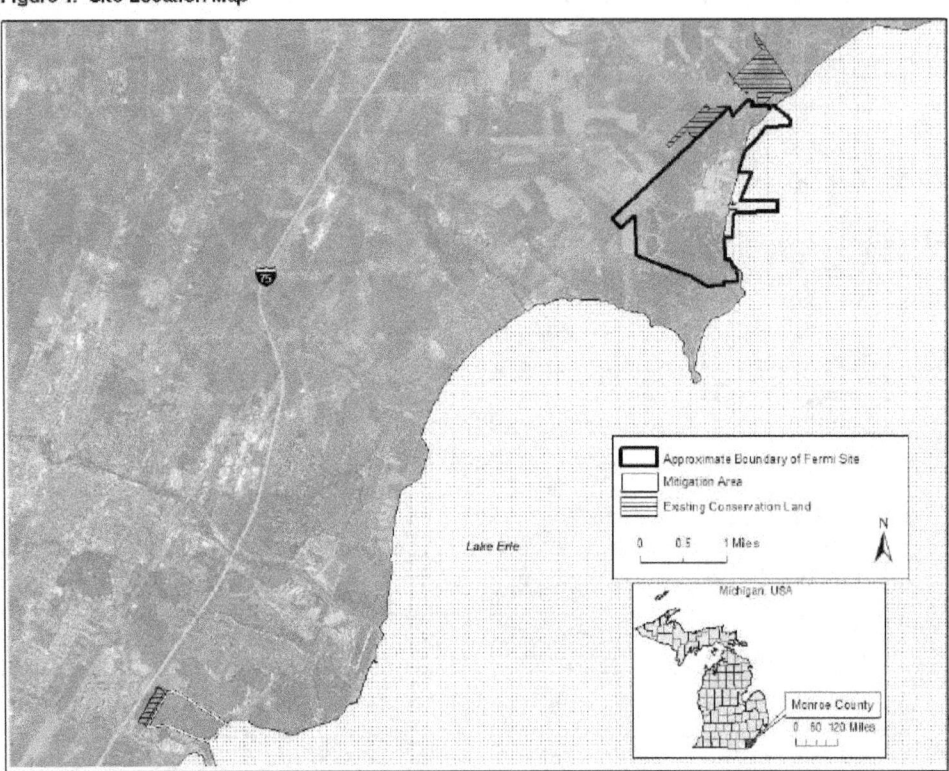

Source: Reference 28

51

Figure 2. Wetland Impact Area Map

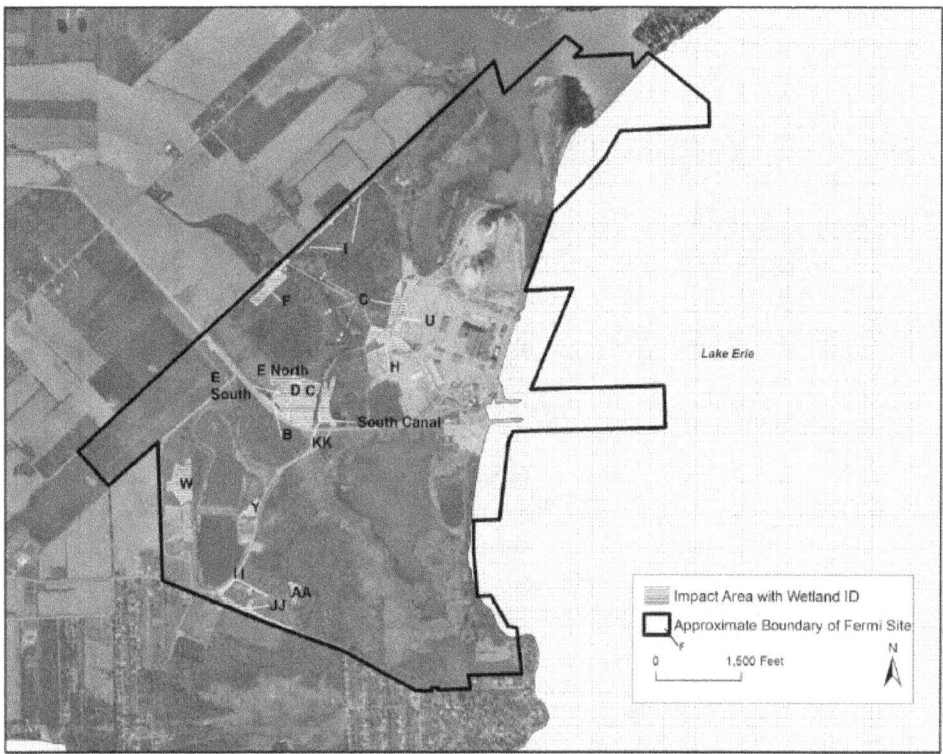

Source: Reference 28

52

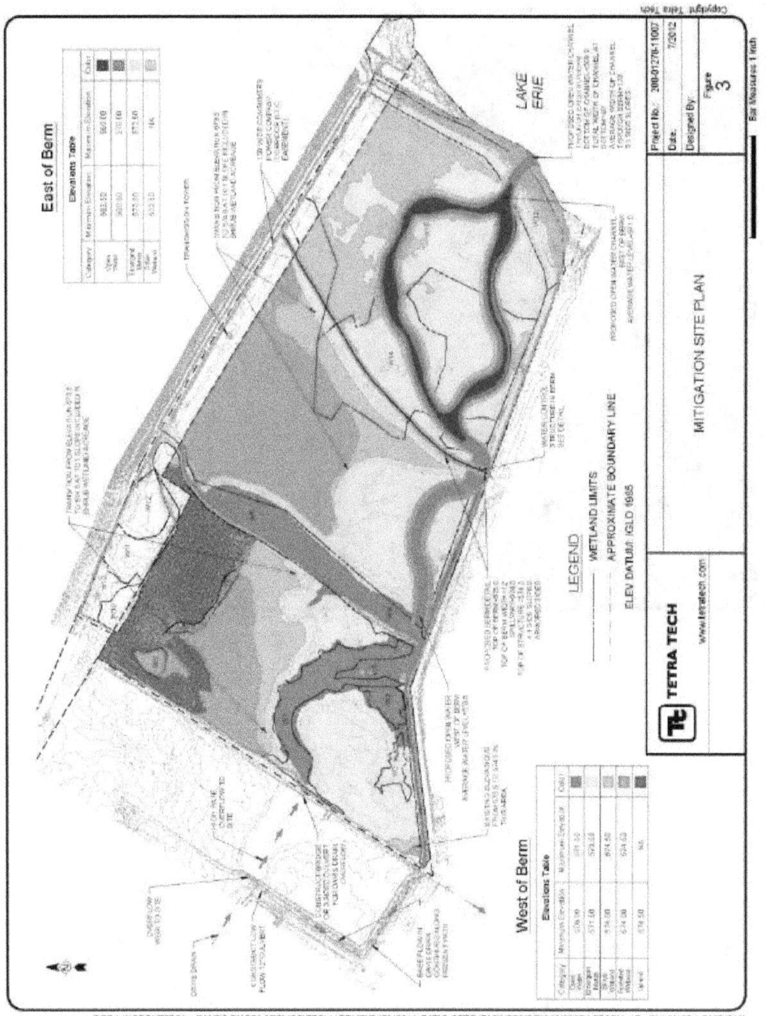

Figure 3. Mitigation Site Plan

Figure 4. Mitigation Acreages

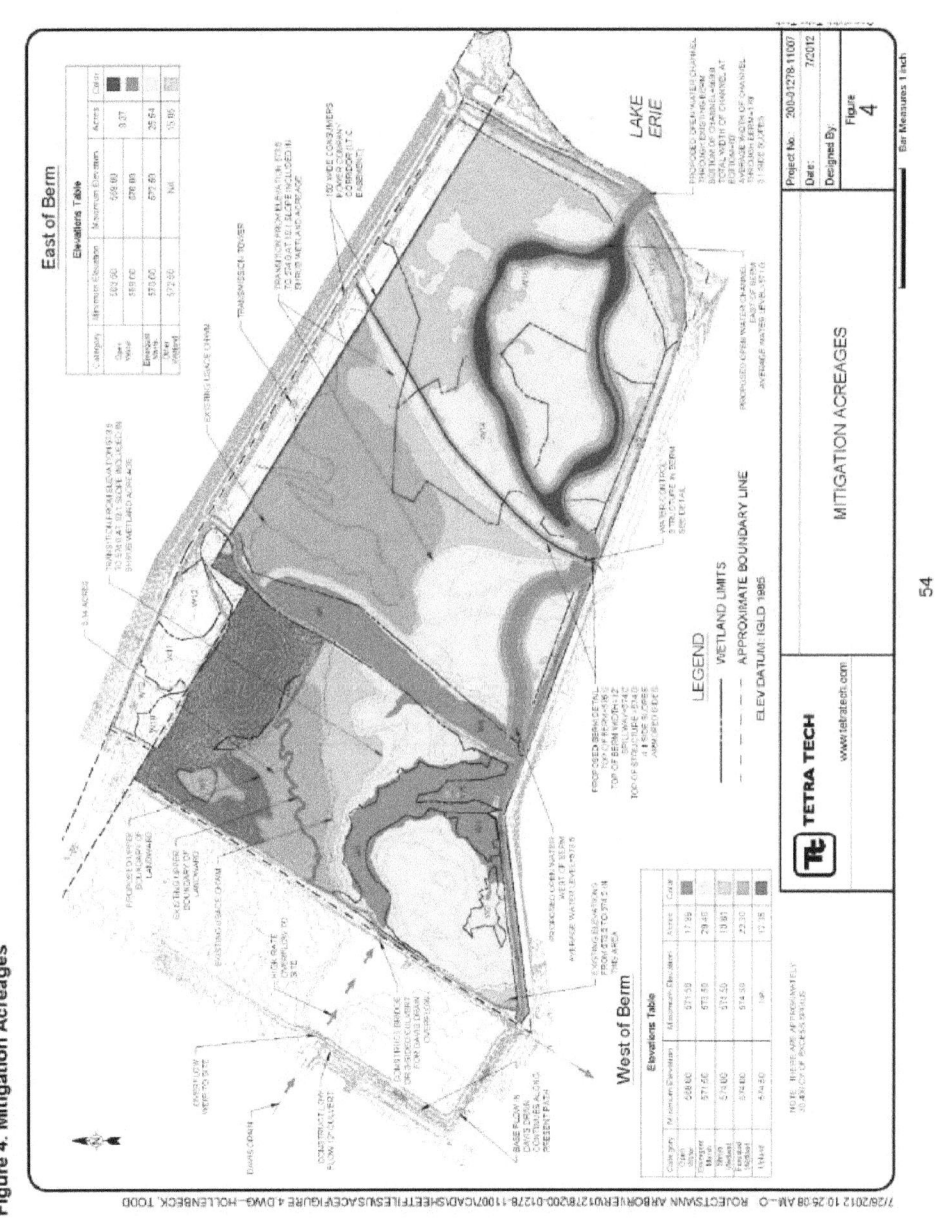

Figure 5. Land Uses on the Fermi Site

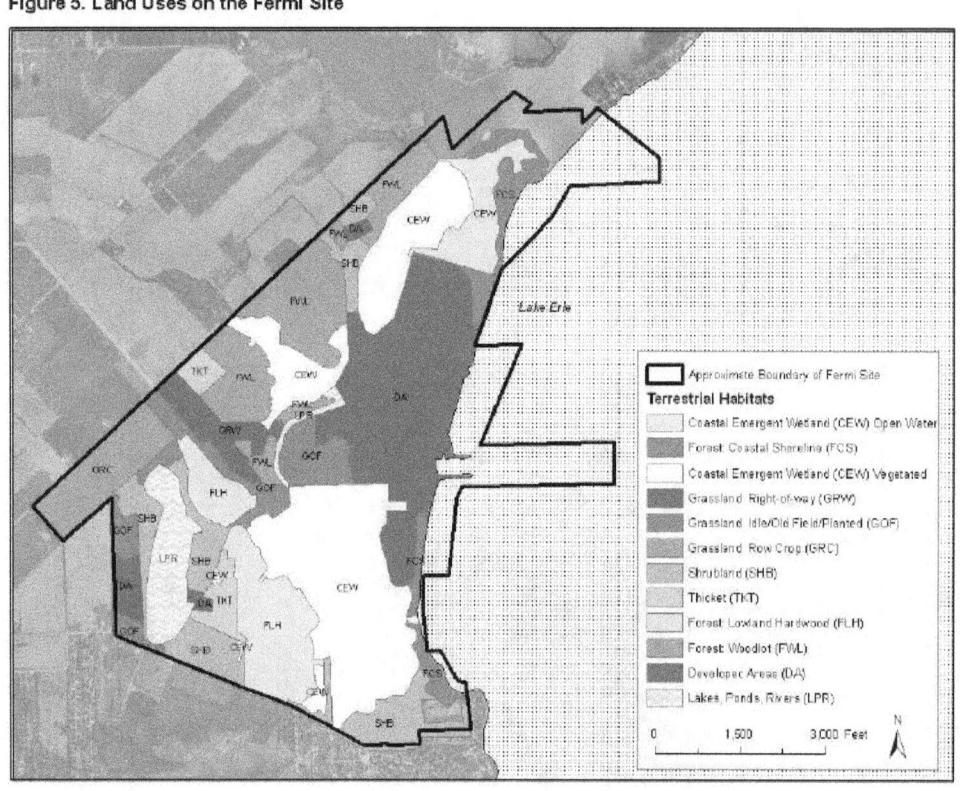

Source: Reference 7

55

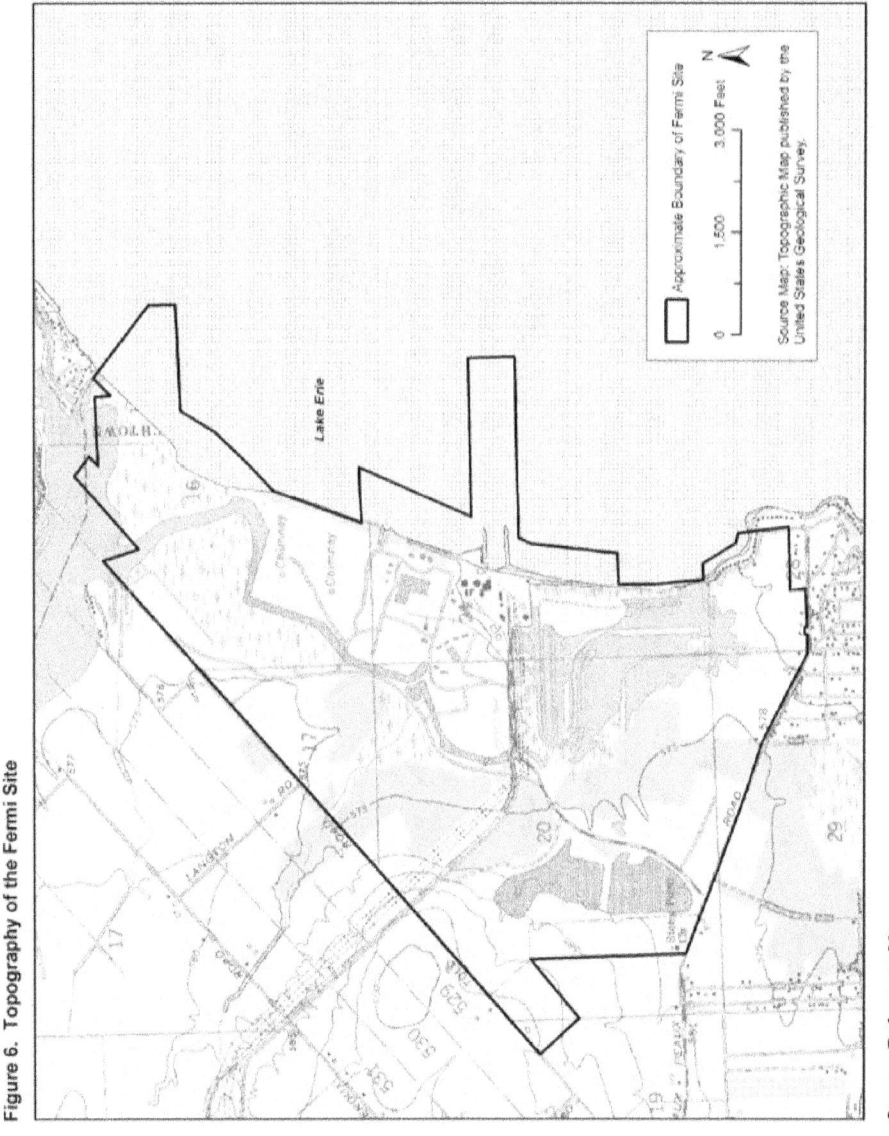

Figure 6. Topography of the Fermi Site

Source Reference 29

56

Figure 7. Soil Types on the Fermi Site

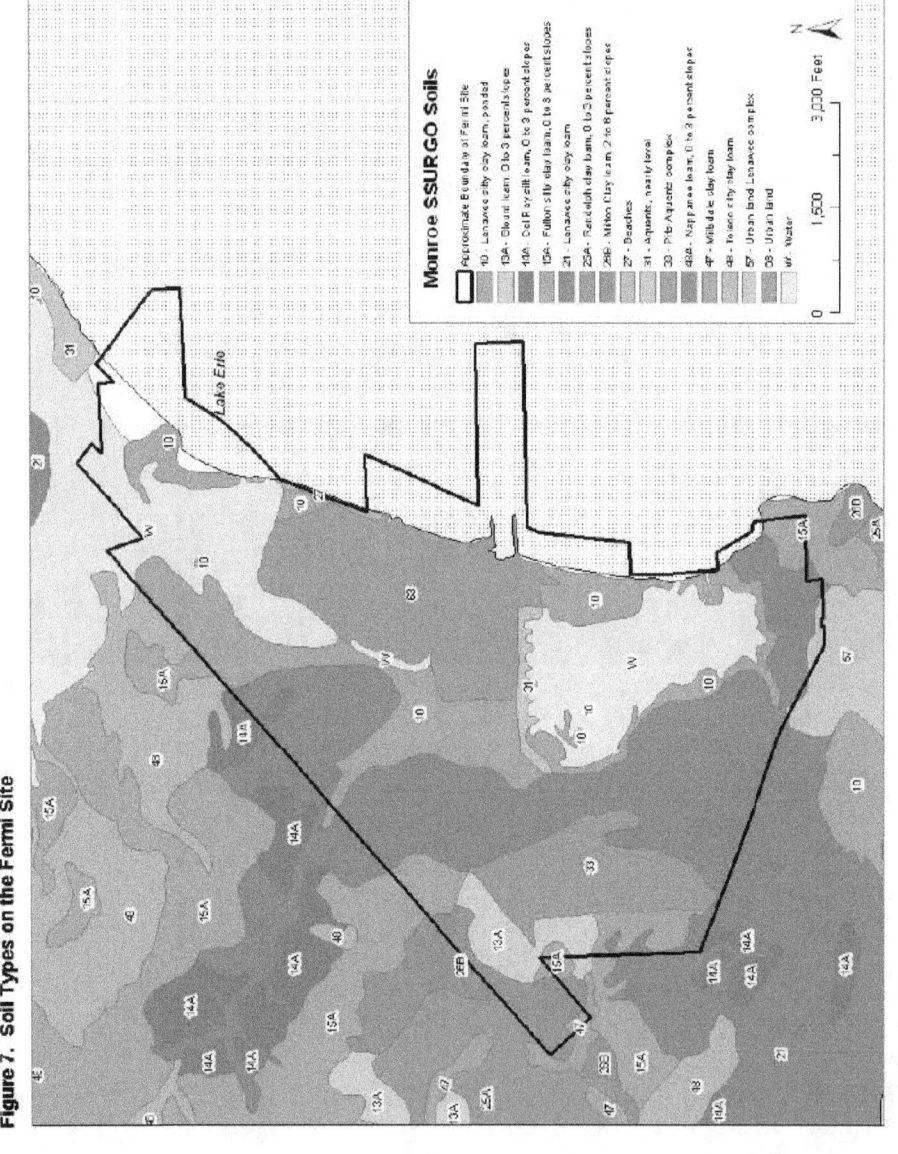

Source: Reference 30

Figure 8. Observed Locations of American Lotus on the Fermi Site

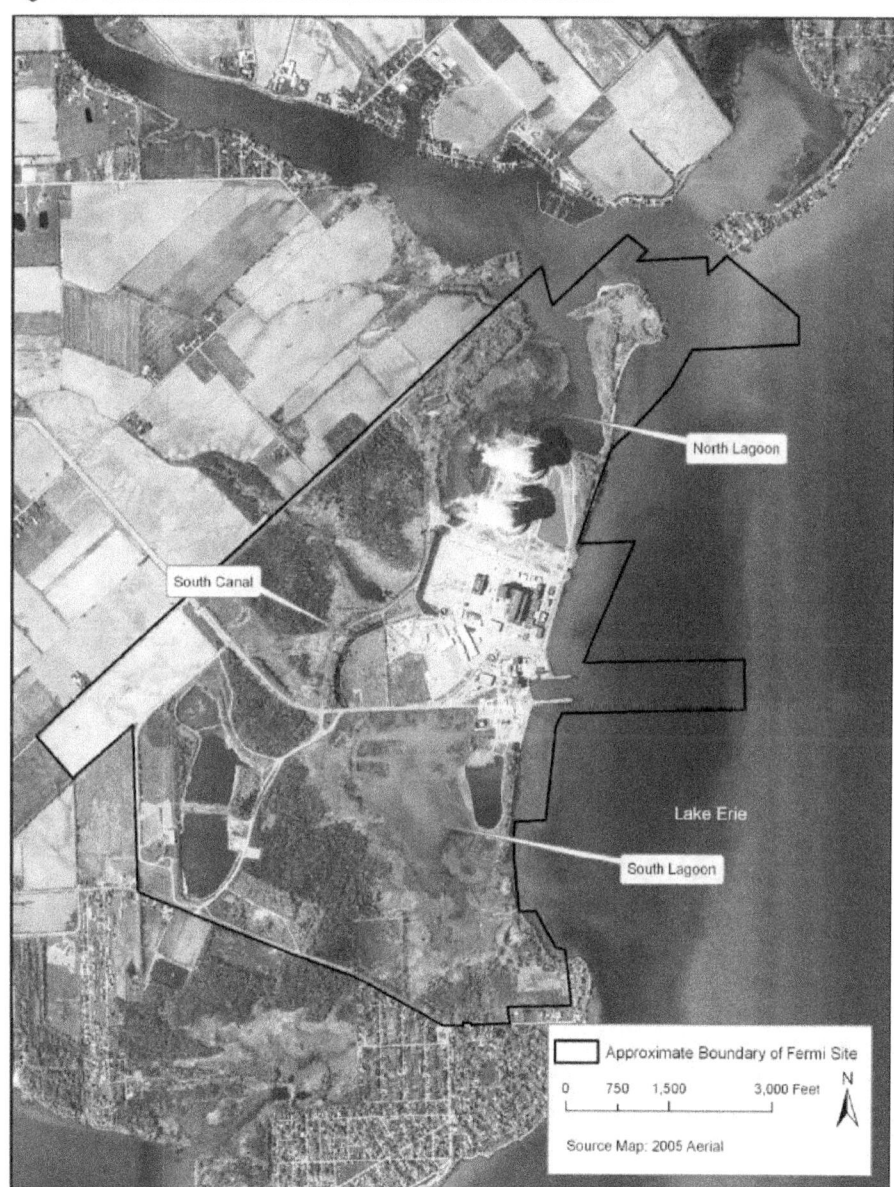

58

Figure 9. Culvert Locations on the Fermi Site

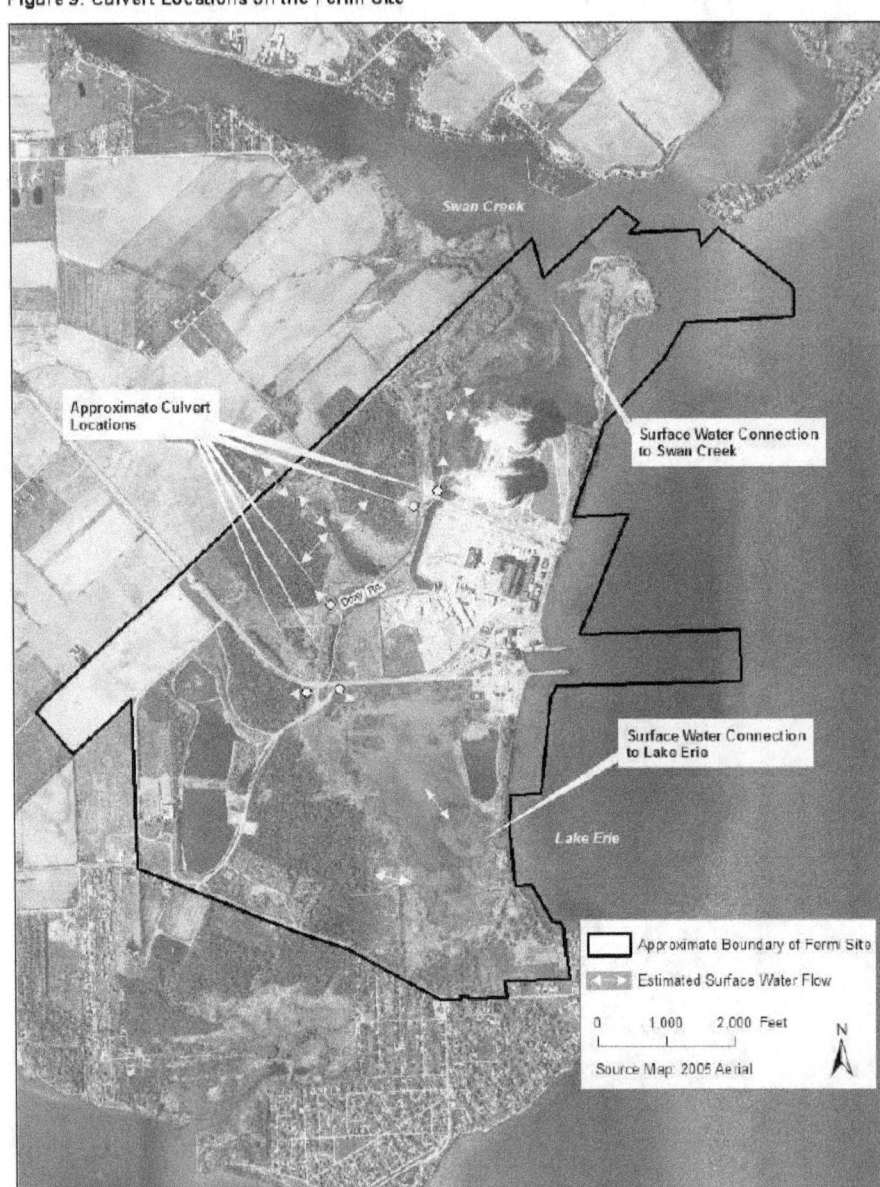

59

Figure 10. Fermi Site Delineated Wetlands

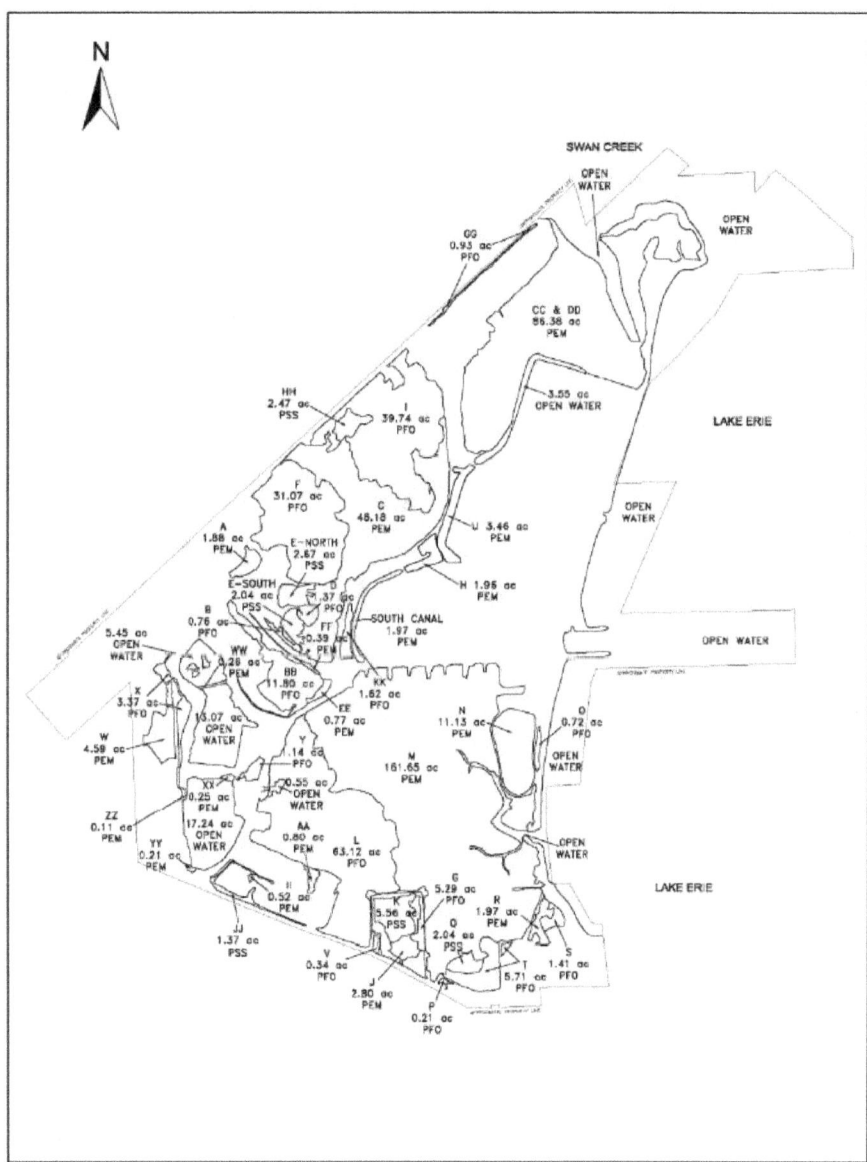

60

Figure 11. Land Use Land Cover (2001) in the Ottawa-Stony Watershed

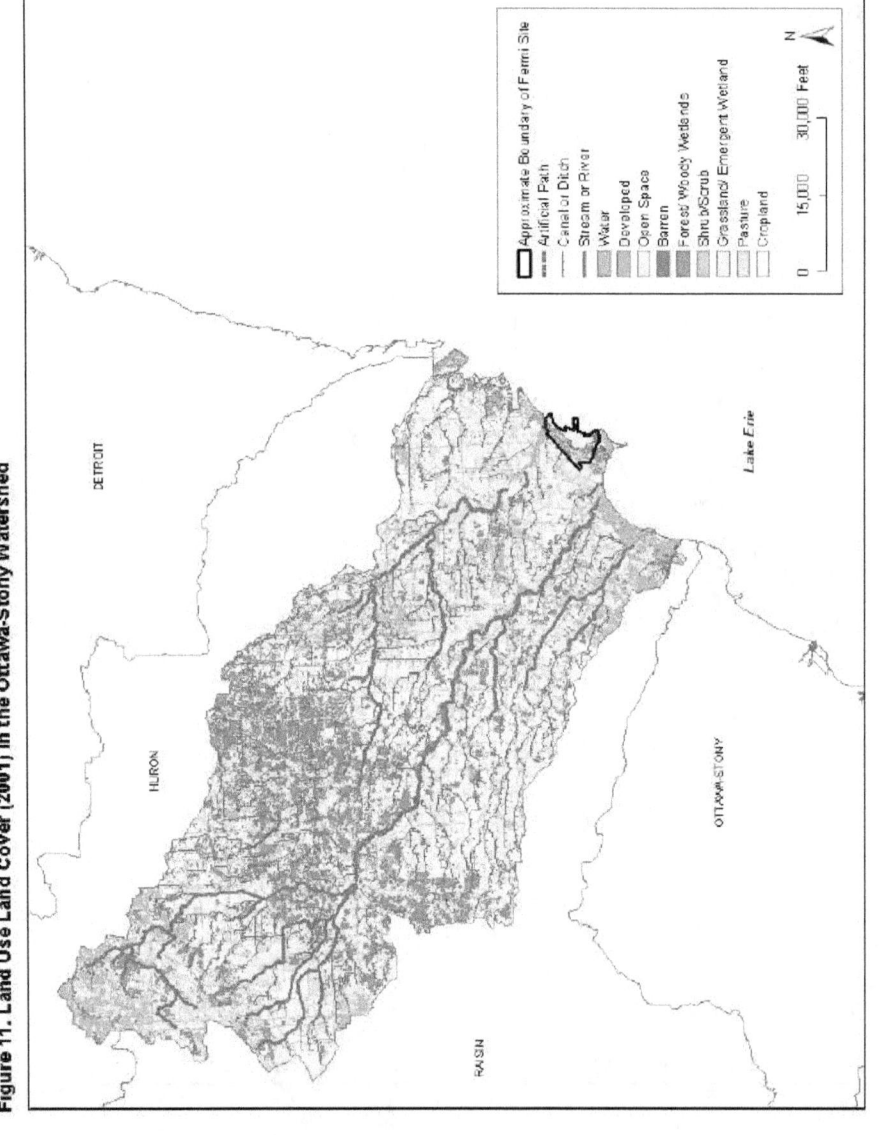

Source: Reference 31 and Reference 32

61

Figure 12. Land Use Land Cover (2001) in the Coastal Zone of Lake Erie

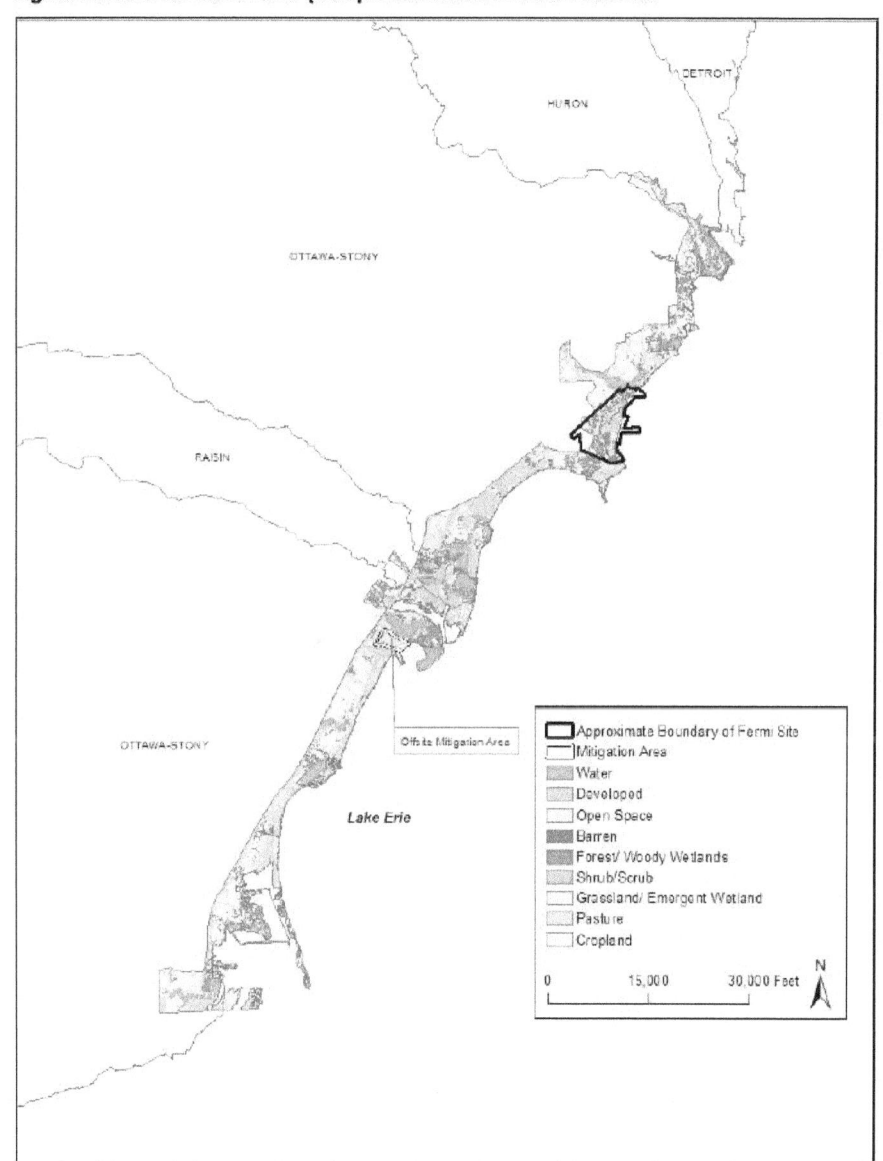

Source: Reference 32 and Reference 33

62

Figure 13. Existing and Former Wetlands in the Ottawa-Stony Watershed

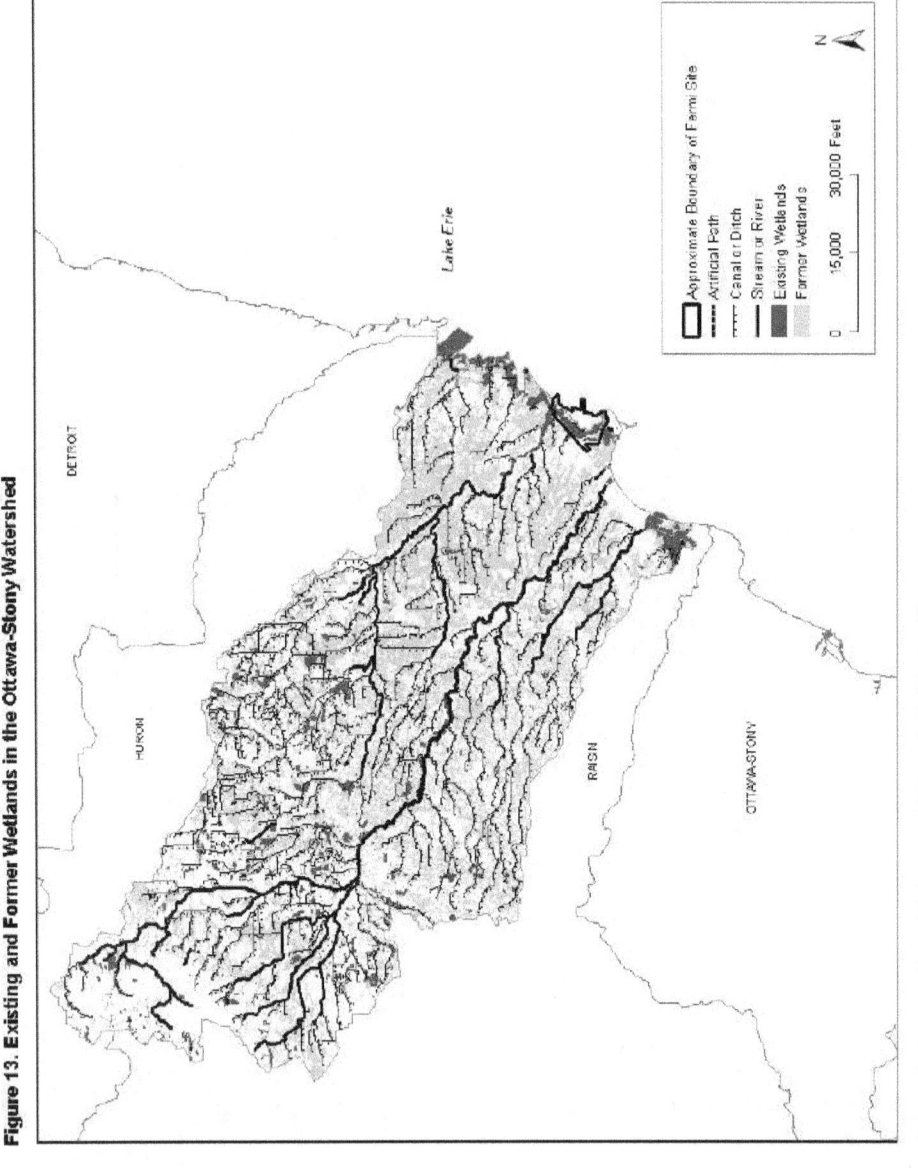

Source: Reference 31 and Reference 34 through Reference 36

63

Figure 14. Existing and Former Wetlands in the Coastal Zone of Lake Erie

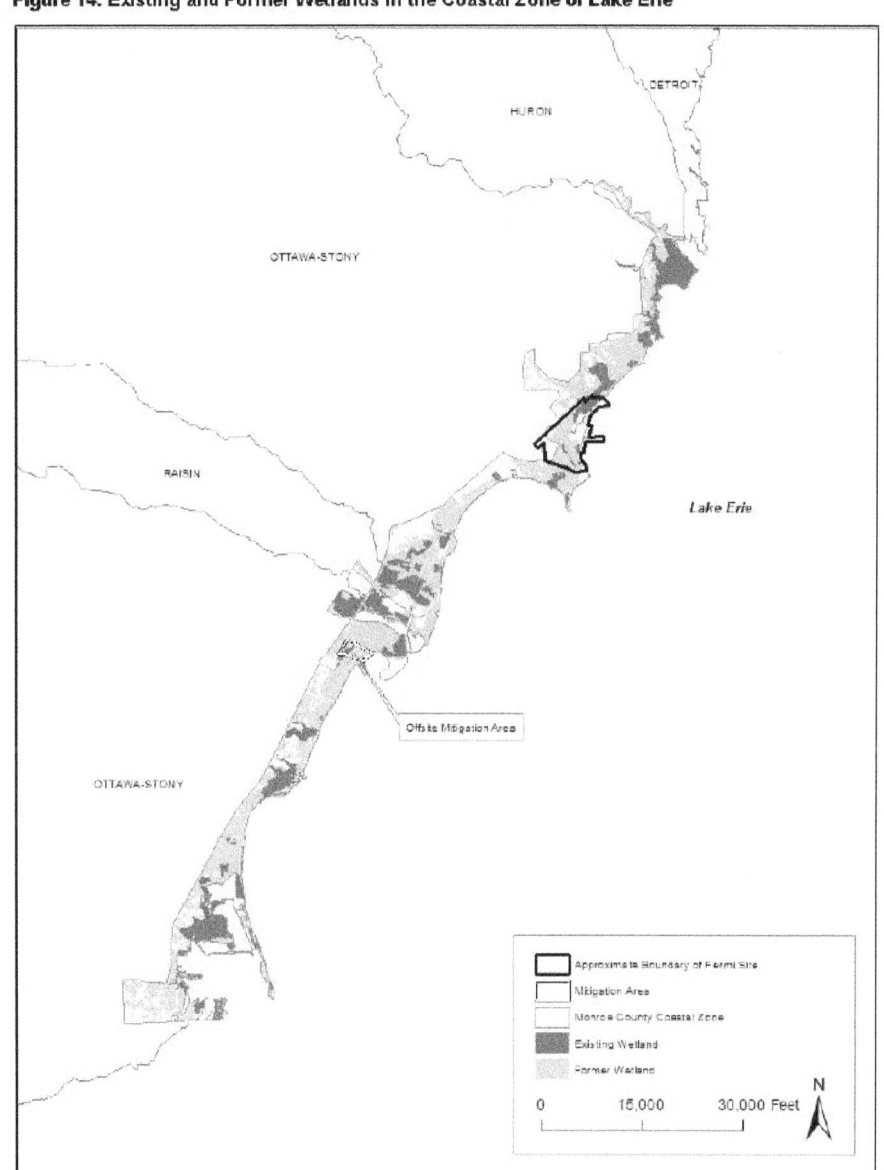

Source: Reference 33 and Reference 36

64

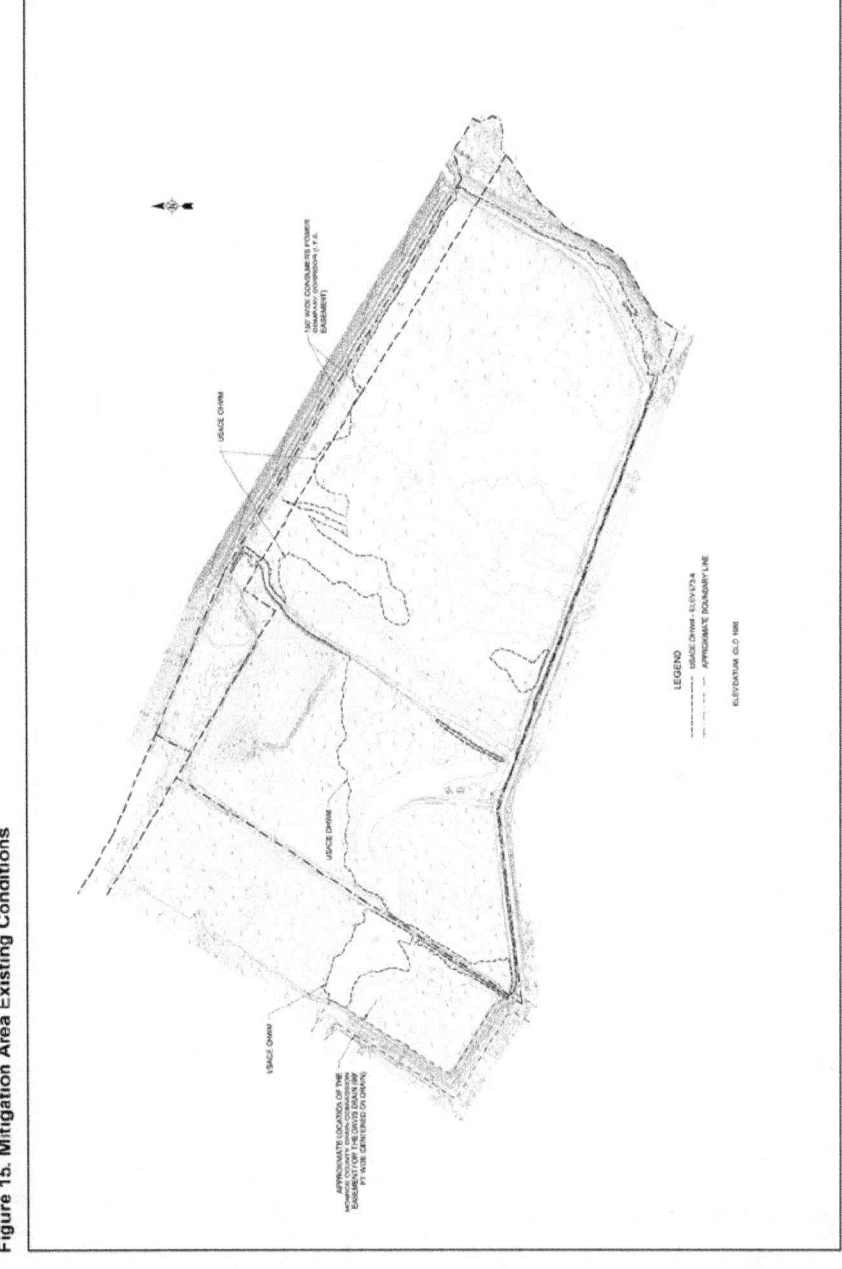

Figure 15. Mitigation Area Existing Conditions

65

Figure 16. Mitigation Area Aerial Photo

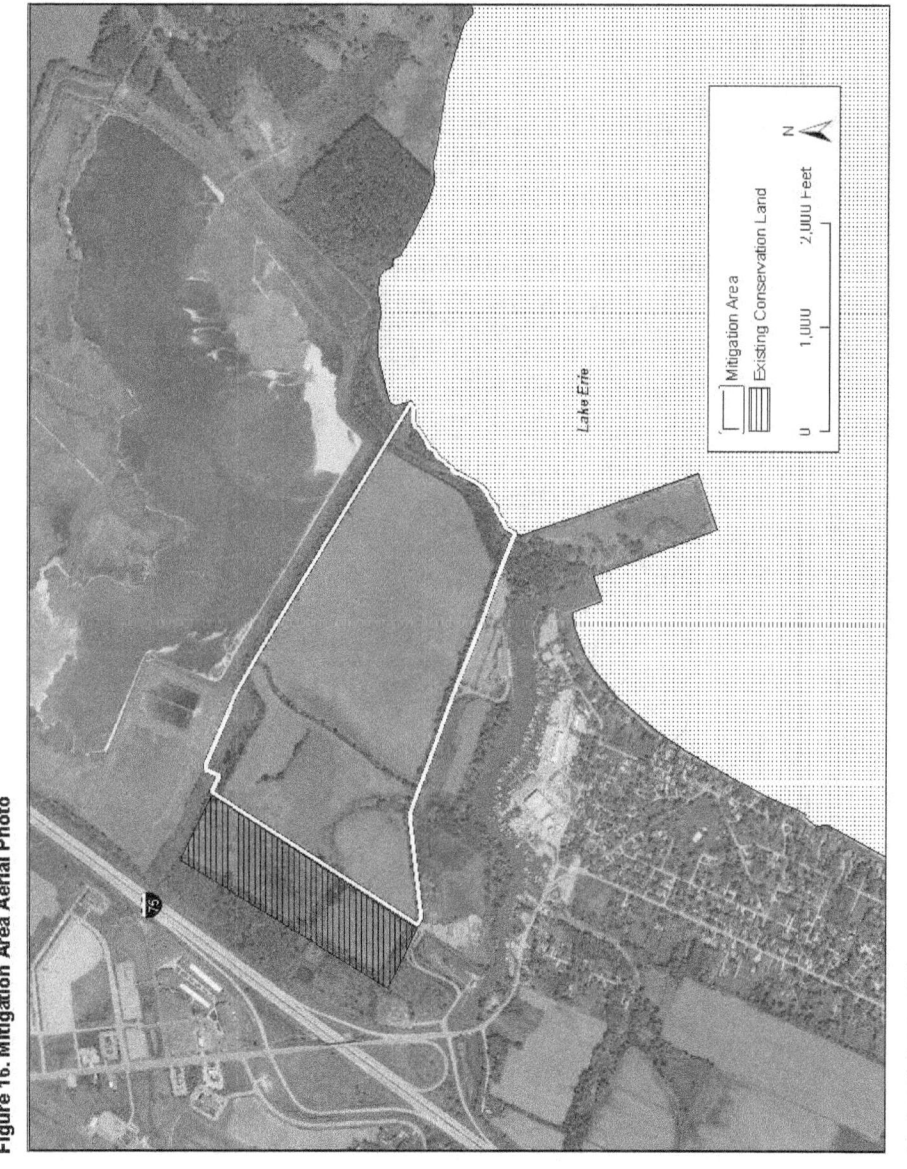

Source: Reference 28

66

Figure 17. Mitigation Area Covertype Map

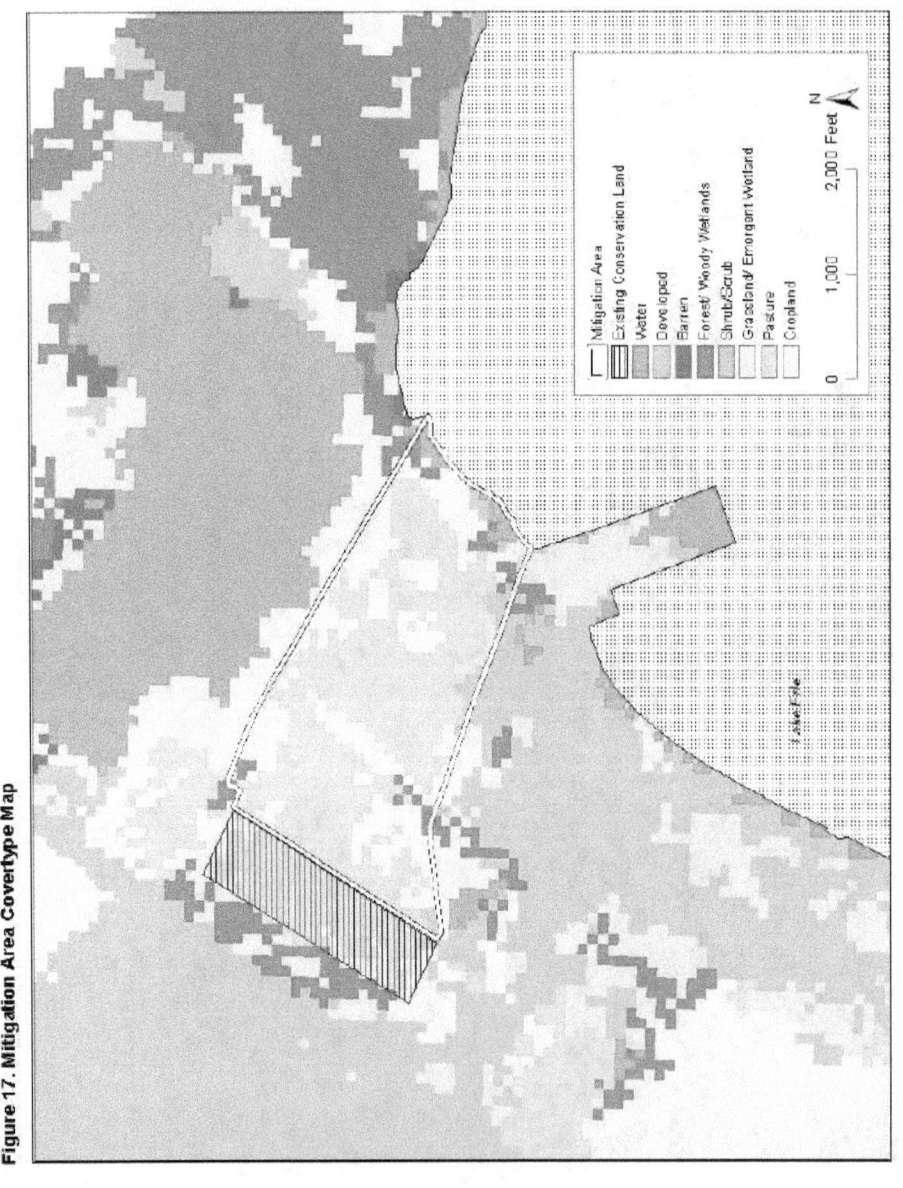

Source: Reference 32

67

Figure 18. Mitigation Area Soils Map

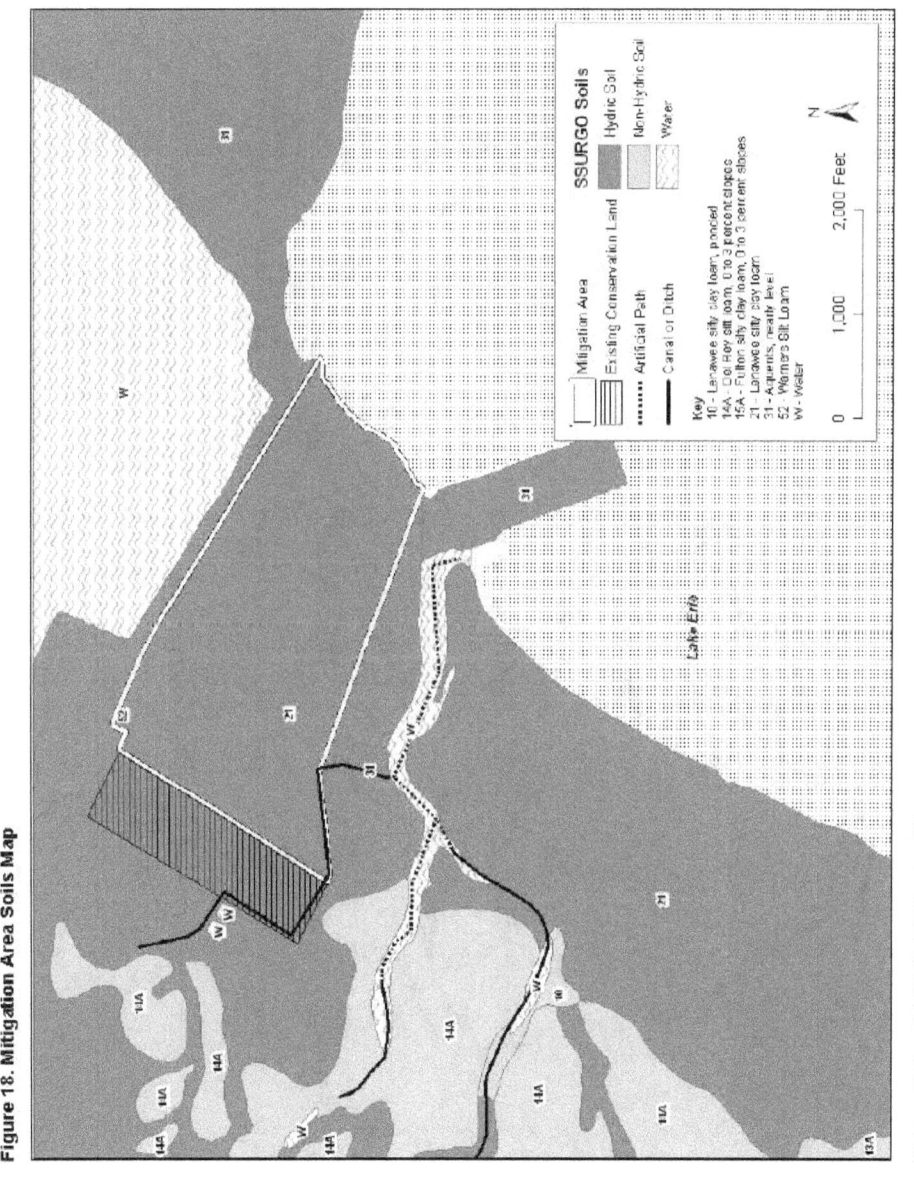

Source: Reference 30 and Reference 31

68

Figure 19. Mitigation Area Current Hydrologic Conditions

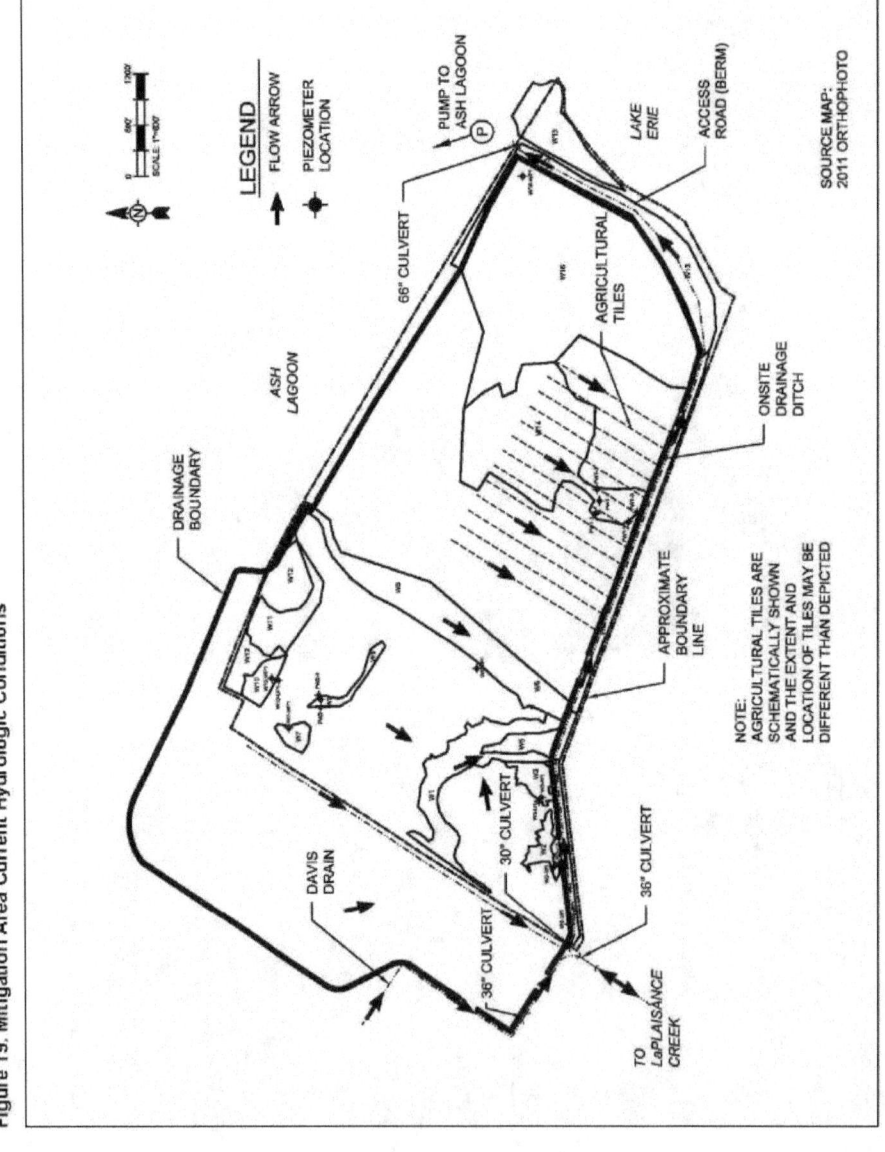

Figure 20. Mitigation Area Federal Mapped Wetlands

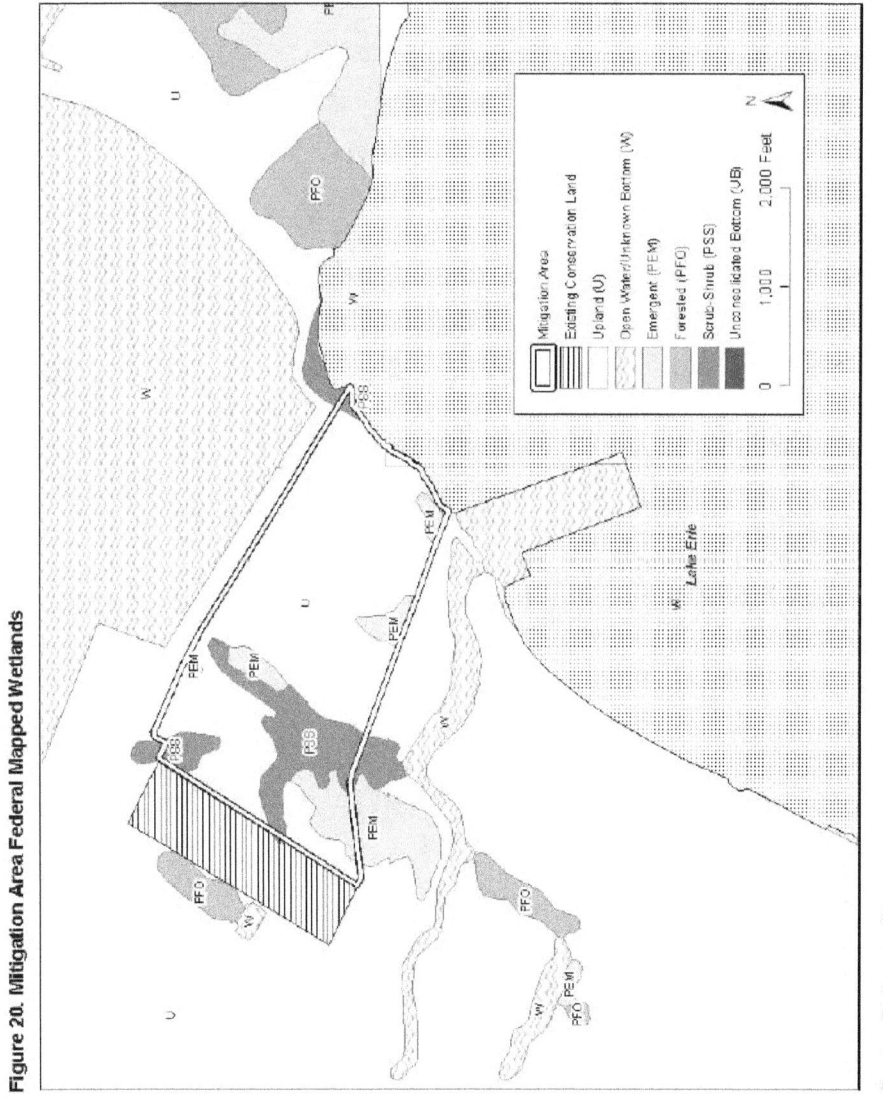

Source: Reference 36

70

Figure 21. Mitigation Area Delineated Wetlands

Figure 22. Mitigation Area Planting Plan

Figure 23. Conservation Easement

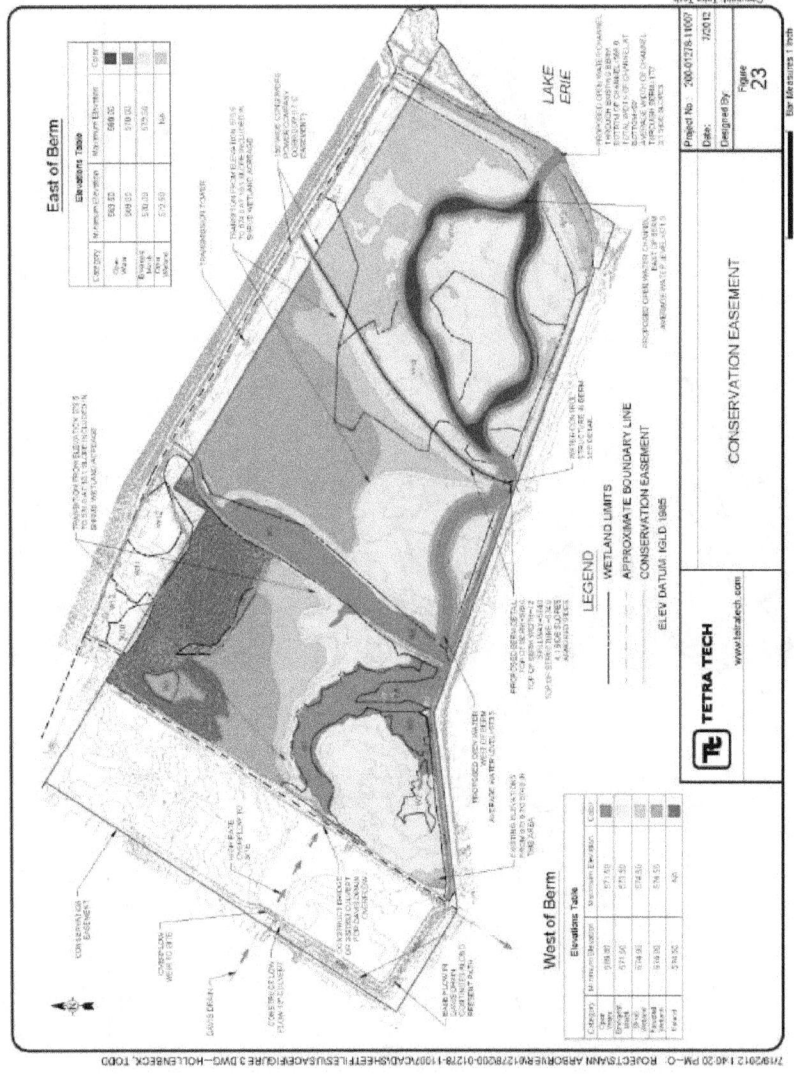

Figure 24. Monitoring Locations

Reference

U.S. Army Corps of Engineers (USACE). 2011. "Public Notice: Proposed Structures and Dredge and Fill Activities Associated with the Proposed Enrico Fermi Unit 3 Nuclear Power Plant in Lake Erie and/or Adjacent Wetlands at Frenchtown Charter Township, Monroe County, Michigan." Accession No. ML12180A374.

Detroit Edison Company (Detroit Edison). 2011. *Detroit Edison Fermi 3 Project, U.S. Army Corps of Engineers and Michigan Department of Environmental Quality, Joint Permit Application.* Revision 1, Detroit Michigan. August. Accession No. ML112700388.

Appendix L

Carbon Dioxide Footprint Estimates for a 1000-MW(e) Light Water Reactor (LWR)

Appendix L

Carbon Dioxide Footprint Estimates for a 1000-MW(e) Light Water Reactor (LWR)

The U.S. Nuclear Regulatory Commission (NRC) review team has estimated the carbon dioxide (CO_2) footprint of various activities associated with nuclear power plants, including building, operating, and decommissioning. The estimates include direct emissions from the nuclear facility and indirect emissions from workforce transportation and the uranium fuel cycle.

Construction equipment estimates listed in Table L-1 are based on hours of equipment use estimated for a single nuclear power plant at a site requiring a moderate amount of terrain modification. A reasonable set of emissions factors used to convert the hours of equipment use to CO_2 emissions is based on carbon monoxide (CO) emissions (UniStar 2007) scaled to CO_2 using a scaling factor of 165 tons of CO_2 per ton of CO. This scaling factor is based on emissions factors in Table 3.3-1 of AP-42 (EPA 1995). Equipment emissions estimated for decommissioning are one-half of those for construction.

Table L-1. Construction Equipment CO_2 Emissions (metric tons equivalent)

Equipment	Construction Total[a]	Decommissioning Total[b]
Earthwork and dewatering	1.1×10^4	5.4×10^3
Batch plant operations	3.3×10^3	1.6×10^3
Concrete	4.0×10^3	2.0×10^3
Lifting and rigging	5.4×10^3	2.7×10^3
Shop fabrication	9.2×10^2	4.6×10^2
Warehouse operations	1.4×10^3	6.8×10^2
Equipment maintenance	9.6×10^3	4.8×10^3
Total[c]	**3.5×10^4**	**1.8×10^4**

(a) Based on hours of equipment usage over 7-year period.
(b) Based on equipment usage over 10-year period.
(c) Total not equal to the sum due to rounding.

Workforce estimates are typical workforce numbers for new plant construction and operation based on estimates in various combined operating license applications; decommissioning workforce emissions estimates are based on decommissioning workforce estimates in NUREG-0586 S1, *Generic Environmental Impact Statement on Decommissioning of Nuclear Facilities, Supplement 1 Regarding the Decommissioning of Nuclear Power Reactors* (NRC 2002). A typical construction workforce averages about 2500 for a 7-year period with a peak workforce of about 4000. A typical operations workforce for the 40-year life of the plant is assumed to be about 400, and the decommissioning workforce during a 10-year decontamination and dismantling period is assumed to be 200 to 400. In all cases, the daily commute is assumed to involve a 100-mi roundtrip with 2 individuals per vehicle. Considering shifts, holidays, and vacations, 1250 roundtrips per day are assumed each day of the year during construction; 200 roundtrips per day are assumed each day during operations; and 150 roundtrips per day are assumed 250 days per year for the decontamination and dismantling portion of decommissioning. If the SAFSTOR decommissioning option is included in decommissioning, 20 roundtrips each day of the year are assumed for the caretaker workforce.

Table L-2 lists the review team's estimates of the CO_2-equivalent emissions associated with workforce transport. The table lists the assumptions used to estimate total miles traveled by each workforce and the factors used to convert total miles to metric tons CO_2-equivalent. The CO_2-equivalent accounts for other greenhouse gases (GHGs), such as methane and nitrous oxide, which are emitted by internal combustion engines. The workers are assumed to travel in gasoline-powered passenger vehicles (cars, trucks, vans, and sport utility vehicles) that get an average of 19.7 mi per gallon of gas (FHWA 2006). Conversion from gallons of gasoline burned to CO_2-equivalent is based on U.S. Environmental Protection Agency (EPA) emissions factors (EPA 2007a, b).

Table L-2. Workforce CO_2 Footprint Estimates

	Construction Workforce	Operational Workforce	Decommissioning Workforce	SAFSTOR Workforce
Roundtrips per day	1250	200	150	20
Miles per roundtrip	100	100	100	100
Days per year	365	365	250	365
Years	7	40	10	40
Miles traveled	3.2×10^8	2.9×10^8	3.8×10^7	2.92×10^7
Miles per gallon[a]	19.7	19.7	19.7	19.7
Gallons fuel burned	1.6×10^7	1.5×10^7	1.9×10^6	1.58×10^6
Metric tons CO_2 per gallon[b]	8.81×10^{-3}	8.81×10^{-3}	8.81×10^{-3}	8.81×10^{-3}
Metric tons CO_2	1.4×10^5	1.3×10^5	1.7×10^4	1.3×10^4
CO_2-equivalent factor[c]	0.971	0.971	0.971	0.971
Metric tons CO_2-equivalent	1.5×10^5	1.3×10^5	1.7×10^4	1.3×10^4

(a) FHWA (2006).
(b) EPA (2007b).
(c) EPA (2007a).

Published estimates of uranium fuel cycle CO_2 emissions required to support a nuclear power plant range from about 1 percent to about 5 percent of the CO_2 emissions from a comparably sized coal-fired plant (Sovacool 2008). A coal-fired power plant emits about 1 metric ton (MT) of CO_2 for each megawatt hour generated (Miller and Van Atten 2004). Therefore, for consistency with Table S-3 of Title 10 of the Code of Federal Regulations (10 CFR 51.51), the NRC staff estimated the uranium fuel cycle CO_2 emissions as 0.05 MT of CO_2 per MWh generated. Finally, the review team estimated the CO_2 emissions directly related to plant operations from the typical usage of various diesel generators onsite using EPA emissions factors (EPA 1995). The review team assumed an average of 600 hr of emergency diesel generator operation per year (total for four generators) and 200 hr of station blackout diesel generator operation per year (total for two generators).

Given the various sources of CO_2 emissions discussed above, the review team estimates the total life CO_2 footprint for a reference 1000-MW(e) nuclear power plant with an 80 percent capacity factor to be about 18 million MT. The components of the footprint are summarized in Table L-3. The uranium fuel cycle component of the footprint dominates all other components. It is directly related to power generated. As a result, it is reasonable to use reactor power to scale the footprint to larger reactors.

Table L-3. 1000-MW(e) LWR Lifetime Carbon Dioxide Footprint

Source	Activity Duration (years)	Total Emissions (metric tons)
Construction equipment	7	3.5×10^4
Construction workforce	7	1.5×10^5
Plant operations	40	1.9×10^5
Operations workforce	40	1.3×10^5
Uranium fuel cycle	40	1.7×10^7
Decommissioning equipment	10	1.8×10^4
Decommissioning workforce	10	1.7×10^4
SAFSTOR workforce	40	1.3×10^4
Total		$\mathbf{1.8 \times 10^7}$

The review team considers the footprint estimated in Table L-3 to be appropriately conservative. The CO_2 emissions estimates for the dominant component (uranium fuel cycle) are based on 30-year-old enrichment technology, assuming that the energy required for enrichment is provided by coal-fired generation. Different assumptions related to the source of energy used for enrichment or the enrichment technology that would be just as reasonable could lead to a significantly reduced footprint.

Emissions estimates presented in the body of this environmental impact statement (EIS) have been scaled to values that are appropriate for the proposed project. The uranium fuel cycle emissions have been scaled by reactor power using the scaling factor determined in Chapter 6. Plant operations emissions have been adjusted to represent the number of large CO_2 emissions sources (diesel generators, boilers, etc.) associated with the project. The workforce emissions estimates have been scaled to account for differences in workforce numbers and commuting distances. Finally, equipment emissions estimates have been scaled by estimated equipment usage. As can be seen in Table L-3, only the scaling of the uranium fuel cycle emissions estimates makes a significant difference in the total carbon footprint of the project.

Sovacool (2008) also calculated GHG emission factors during the life cycle of nuclear power plants based on the statistical analysis from 19 qualified studies examined. Estimated GHG emission factors ranged from 1.4 g CO_2-equivalent per kWh to 288 g CO_2-equivalent per kWh, with a mean value of 66 g CO_2-equivalent per kWh (equivalent to 0.066 MT of CO_2-equivalent per MWh). The emission factor of 0.05 MT of CO_2 per MWh used in this analysis is about three-fourths the mean emission factor of 0.066 MT of CO_2-equivalent per MWh but is considered comparable, considering the wide range of emission factors (0.0014 to 0.288) estimated in that study.

L.1   References

10 CFR Part 51. Code of Federal Regulations, Title 10, *Energy*, Part 51, "Environmental Protection Regulations for Domestic Licensing and Related Regulatory Functions."

Federal Highway Administration (FHWA). 2006. *Highway Statistics 2005* (Table VM-1). Office of Highway Policy Information, Washington, D.C.

Miller, P.J., and C. Van Atten. 2004. *North American Power Plant Air Emissions*. Commission for Environmental Cooperation of North America, Montreal.

Sovacool, B.K. 2008. "Valuing the Greenhouse Gas Emissions from Nuclear Power: A Critical Survey." *Energy Policy* 36:2940–2953.

UniStar Nuclear Energy, LLC (UniStar). 2007. *Technical Report in Support of Application of UniStar Nuclear Energy, LLC and UniStar Nuclear Operating Services, LLC, for Certificate of Public Convenience and Necessity before the Maryland Public Service Commission for Authorization to Construct Unit 3 at Calvert Cliffs Nuclear Power Plant and Associated Transmission Lines*. Prepared for the Public Service Commission of Maryland, dated November 6, 2007. Accession No. ML090680065.

U.S. Environmental Protection Agency (EPA). 1995. *Compilation of Air Pollutant Emission Factors, Volume 1: Stationary and Point and Area Sources*. AP-42, 5th Edition, Office of Air and Radiation, Research Triangle Park, North Carolina.

U.S. Environmental Protection Agency (EPA). 2007a. *Inventory of U.S. Greenhouse Gas Emissions and Sinks: 1990–2005* (Table 3-7). Washington, D.C.

U.S. Environmental Protection Agency (EPA). 2007b. "Conversion Factors to Energy Units (Heat Equivalents) Heat Contents and Carbon Content Coefficients of Various Fuel Types" in: *U.S. Inventory of Greenhouse Gas Emissions and Sinks 1990–2005.* EPA 430-F-07-004, Washington, D.C.

U.S. Nuclear Regulatory Commission (NRC). 2002. *Generic Environmental Impact Statement on Decommissioning of Nuclear Facilities, Supplement 1 Regarding the Decommissioning of Nuclear Power Reactors.* NUREG-0586 S1, Volume 1, Washington, D.C.

Appendix M

Environmental Impacts from Building and Operating Transmission Lines Proposed to Serve Fermi 3

Appendix M

Environmental Impacts from Building and Operating Transmission Lines Proposed to Serve Fermi 3

The final environmental impact statement (EIS) presents integrated evaluations of potential environmental impacts from the proposed Fermi 3 facilities, organized by environmental resource. The review team's evaluation of potential environmental impacts from building and operating electrical transmission lines that may be built to serve the proposed Fermi 3 facility is found in those places in the final EIS text that address environmental resources that would be affected by the proposed transmission lines. Offsite transmission lines are not part of the Fermi 3 COL application, and any such lines would be built by ITC*Transmission* rather than Detroit Edison. Under NRC regulations in 10 CFR 50.10(a)(2)(vii), building of transmission lines is a preconstruction activity not subject to the Nuclear Regulatory Commission's regulatory authority. However, many preconstruction activities are within the regulatory authority of local, State, or other Federal agencies, and certain preconstruction activities require a permit from the U.S. Army Corps of Engineers.

This appendix provides a brief roadmap to where in the final EIS environmental impacts from transmission lines are addressed. In the final EIS, the environmental impacts of transmission lines are primarily described in terms of the following resource areas: (1) land use, (2) terrestrial ecology, (3) aquatic ecology, (4) historical and cultural resources, and (5) nonradiological health. The proposed route for the new transmission lines is described in Section 3.2.2.3 and shown in Figure 3-8. Table M-1 lists the sections/subsections of Chapter 2 (Affected Environment), Chapter 4 (Construction Impacts at the Proposed Site), Chapter 5 (Operational Impacts at the Proposed Site), and Chapter 7 (Cumulative Impacts) that contain pertinent information related to the review team's evaluation of potential impacts from the transmission lines.

The review team considered transmission line impacts for all environmental resource areas addressed in Chapters 2, 3, 4, 5, and 7, not just those resources highlighted in Table M-1. However, the discussion for other resources is limited in the final EIS text because construction and operation of transmission lines have limited relevance to impacts on these resource areas.

Table M-1. Sections of the EIS in Which Potential Impacts from Transmission Lines Are Discussed

Resource Area	Affected Environment	Construction and Preconstruction Impacts	Operations Impacts	Cumulative Impacts
Land Use	2.2.2	4.1.2	5.1.2	7.1[a]
Terrestrial Ecology	2.4.1.2	4.3.1.2	5.3.1.2	7.3.1[a]
Aquatic Ecology	2.4.2.2	4.3.1.2	5.3.2.2	7.3.2[a]
Historic and Cultural Resources	2.7.3	4.6.2	5.6[a]	7.5[a]
Nonradiological Health	2.10.4	4.8.1.2[a]	5.8.3, 5.8.4	7.7[a]
Summaries/Conclusions	Figure 2-5, Table 2-9, Table 2-63	Table 4-22, Table 4-23	Table 5-35, Table 5-36	Table 7-3[b]

(a) Only certain parts of the indicated sections are specifically focused on transmission lines.

(b) Although Table 7-3 does not specifically mention transmission lines, the conclusions presented in the table account for transmission line impacts.

In addition, the review team considered the potential impacts of building and operating transmission lines associated with the use of each of the four alternative plant sites evaluated in Sections 9.3.3, 9.3.4, 9.3.5, and 9.3.6. The final conclusions and recommendations, summarized in Chapter 10 and in Tables 10-1, 10-2, and 10-4, regarding environmental impacts for the overall Fermi 3 project also account for potential transmission line impacts.

NRC FORM 335
(12-2010)
NRCMD 3.7

U.S. NUCLEAR REGULATORY COMMISSION

BIBLIOGRAPHIC DATA SHEET

(See instructions on the reverse)

1. REPORT NUMBER (Assigned by NRC, Add Vol., Supp., Rev., and Addendum Numbers, If any.)
NUREG 2105, Vol. 4

2. TITLE AND SUBTITLE

Environmental Impact Statement for Combined License (COL) for Enrico Fermi Unit 3
Final Report

3. DATE REPORT PUBLISHED	
MONTH	YEAR
January	2013

4. FIN OR GRANT NUMBER

5. AUTHOR(S)

See Appendix A

6. TYPE OF REPORT

Technical

7. PERIOD COVERED (Inclusive Dates)

8. PERFORMING ORGANIZATION - NAME AND ADDRESS (If NRC, provide Division, Office or Region, U. S. Nuclear Regulatory Commission, and mailing address; if contractor, provide name and mailing address.)

Division of New Reactor Licensing
Office of New Reactors
U. S. Nuclear Regulatory Commission
Washington, D.C. 20555-0001

9. SPONSORING ORGANIZATION - NAME AND ADDRESS (If NRC, type "Same as above", if contractor, provide NRC Division, Office or Region, U. S. Nuclear Regulatory Commission, and mailing address.)

Same as above

10. SUPPLEMENTARY NOTES

Docket No. 52-033

11. ABSTRACT (200 words or less)

This environmental impact statement (EIS) has been prepared in response to an application submitted to the U.S. Nuclear Regulatory Commission (NRC) by Detroit Edison for a construction permit and operating license (combined license or COL). The proposed actions related to the Detroit Edison application are (1) NRC issuance of a COL for a new power reactor unit at the Detroit Edison Enrico Fermi Atomic Power Plant (Fermi) site in Monroe County, Michigan; and (2) U.S. Army Corps of Engineers (USACE) permit action to perform certain regulated activities on the site. The USACE is participating with the NRC in preparing this EIS as a cooperating agency and participates collaboratively on the review team.

After considering the environmental aspects of the proposed action, the staff's recommendation to the Commission is that the COL be issued as proposed. This recommendation is based on (1) the application, including the Environmental Report (ER) submitted by Detroit Edison; (2) consultation with Federal, State, Tribal, and local agencies; (3) the staff's independent review; (4) the staff's consideration of comments related to the environmental review that were received during the public scoping process and on the draft EIS; and (5) the assessments summarized in this EIS, including the potential mitigation measures identified in the ER and this EIS. The USACE permit decision would be made following issuance of this final EIS and completion of its permit application review process and permit decision documentation.

12. KEY WORDS/DESCRIPTORS (List words or phrases that will assist researchers in locating the report.)

Enrico Fermi Unit 3
National Environmental Policy Act, NEPA
Final Environmental Impact Statement, FEIS
Combined License, COL, COLA
Environmental Review
New Reactors

13. AVAILABILITY STATEMENT
unlimited

14. SECURITY CLASSIFICATION
(This Page)
unclassified
(This Report)
unclassified

15. NUMBER OF PAGES

16. PRICE

Federal Recycling Program

UNITED STATES
NUCLEAR REGULATORY COMMISSION
WASHINGTON, DC 20555-0001

OFFICIAL BUSINESS

NUREG-2105
Volume 4, Final

Environmental Impact Statement for the Combined License (COL)
for Enrico Fermi Unit 3

January 2013

www.ingramcontent.com/pod-product-compliance
Lightning Source LLC
Chambersburg PA
CBHW080235180526
45167CB00006B/2289